Stochastic Population Dynamics in Ecology and Conservation

Oxford Series in Ecology and Evolution
Edited by Paul H. Harvey and Robert M. May

Stochastic Population Dynamics in Ecology and Conservation

RUSSELL LANDE
University of California, San Diego

STEINAR ENGEN
Norwegian University of Science and Technology

BERNT-ERIK SÆTHER
Norwegian University of Science and Technology

OXFORD
UNIVERSITY PRESS

OXFORD

UNIVERSITY PRESS

Great Clarendon Street, Oxford OX2 6DP

Oxford University Press is a department of the University of Oxford.
It furthers the University's objective of excellence in research, scholarship,
and education by publishing worldwide in

Oxford New York

Auckland Bangkok Buenos Aires Cape Town Chennai
Dar es Salaam Delhi Hong Kong Istanbul Karachi Kolkata
Kuala Lumpur Madrid Melbourne Mexico City Mumbai Nairobi
São Paulo Shanghai Taipei Tokyo Toronto

Oxford is a registered trade mark of Oxford University Press
in the UK and in certain other countries

Published in the United States
by Oxford University Press Inc., New York

A catalogue record for this title is available from the British Library

Library of Congress Cataloging in Publication Data
Lande, Russell.
Stochastic population dynamics in ecology and conservation / Russell Lande,
Steinar Engen, Bernt-Erik Sæther.
(Oxford series in ecology and evolution)
Includes bibliographical references (p.).
1. Population biology–Mathematical models. 2. Ecology–Mathematical models.
3. Conservation biology–Mathematical models. 4. Stochastic–processes–Mathametical
models. I. Engen, Steinar II. Sæther, Bernt-Erik III. Title IV. Series.

QH352 .L36 2003   577.8′8′01518–dc21   2002042559

ISBN 0 19 852524 9 (hbk.)

ISBN 0 19 852525 7 (pbk.)

10 9 8 7 6 5 4 3 2 1

Typeset by Newgen Imaging Systems (P) Ltd., Chennai, India
Printed in Great Britain
on acid-free paper by Biddles Ltd, Guildford & King's Lynn

Contents

8 Community dynamics 162

References 182

Index 209

Preface

All natural populations fluctuate in a substantially unpredictable, stochastic fashion. Stochasticity creates a risk of population extinction that does not exist in deterministic models, with important consequences for both pure and applied ecology. This book introduces biological concepts and mathematical methods that have proven useful in analyzing stochastic population dynamics in time and space. Assuming the reader has a background in basic ecology and a working knowledge of elementary calculus and statistics, we emphasize the insights that can be gained from relatively simple analytical models that incorporate stochasticity.

We make no attempt at a complete coverage of the field in terms of either taxa or the range of possible population models. For applications of single population models, we have a taxonomic preference for terrestrial vertebrates, particularly birds and large mammals, for two reasons. First, the most accurate data, with complete population censuses over decades, are available for birds and to a lesser extent for mammals. This facilitates our focus on biological processes in stochastic population dynamics without devoting inordinate space to statistical methods for estimating population sizes and handling large uncertainty that characterizes most population estimates. Second, many vertebrate populations reproduce annually at a certain breeding season each year, with small maximum rates of population growth. This greatly simplifies their analysis by avoiding complications of cyclic and chaotic dynamics that may apply to populations with short generations and high growth rates. Species diversity and community dynamics are illustrated using data on insects, especially tropical butterflies that are taxonomically well known and amenable to standardized collection of large samples, including numerous species over multiple generations.

In Chapter 1 we define and formulate demographic and environmental stochasticity, and illustrate statistical methods for estimating them from field data. The concept of long-run growth rate of a population in a stochastic environment is explained and, in Chapter 3, extended to age-structured populations. Chapters 2 and 4 develop diffusion approximations to analyze the extinction dynamics of single populations and metapopulations with local demographic and environmental stochasticity. Chapter 3 presents a new method of estimating the total density dependence in a life history from time series of populations with discrete annual reproduction. Applications to several data sets confirm that vertebrate population dynamics usually are undercompensated and not intrinsically cyclic or chaotic. Chapter 4 derives a model of small or moderate fluctuations in populations with a continuous spatial distribution to analyze the spatial scale of population fluctuations and local extinction risk. In Chapter 5, stochastic dynamics and statistical uncertainty are incorporated into population viability analysis using population prediction intervals. Chapter 6 demonstrates that stochasticity and uncertainty can strongly affect sustainable harvesting strategies. Statistics of species diversity measures and species abundance distributions are summarized in Chapter 7, with implications for the reliability of rapid assessments of biodiversity. Partitioning diversity into additive components associated with different factors, similar to the analysis of variance, provides a new approach to analyzing patterns of species diversity. In Chapter 8, we derive stochastic community models to analyze spatio-temporal data on community dynamics, revealing that the great majority of the variance of logarithmic species abundances can be attributed to deterministic and stochastic factors absent from neutral community models.

Although the conceptual simplicity and potential realism of simulation models appeals to many researchers, there are serious limitations and possible pitfalls that an analytically unsophisticated researcher is likely to encounter. Computer simulation of realistic models is not conducive to reaching general conclusions, and may also be time consuming, because each simulation must necessarily be based on particular values of various parameters. To paraphrase the eminent twentieth century British biologist J. B. S. Haldane, an ounce of analysis is worth a ton of simulation. Statistical theory has established that the best predictive power is achieved by relatively simple models with a small or intermediate number of parameters, rather than by complex models, because limited data do not permit accurate estimation of many parameters. A theme repeated throughout this book is that analytical modeling or insight derived from analysis, combined with simulation, often provides the most informative approach for understanding and interpreting population data.

For access to unpublished data and preprints we thank the British Trust for Ornithology, Nicholas J. Aebischer, Frank Adriaensen, Tim Clutton-Brock, Tim Coulson, Thomas O. Crist, Brian Dennis, Philip J. DeVries, Flurin Filli, Erik Matthysen, Robin H. McCleery, Chris Perrins, Rolf Peterson, Joseph A. Veech and Henri Weimerskirch. Constructive criticisms were provided by Ted J. Case and Peter Kareiva (all chapters), Kaustuv Roy (Chapter 2), and Philip J. DeVries (Chapters 7, 8). We thank Ingunn Yttersian for help in preparing the manuscript. This work was partially supported by grants from the U.S. National Science Foundation, and the Research Council of Norway.

1

Demographic and environmental stochasticity

1.1 Stochastic population fluctuations

Fluctuations in population size often appear to be stochastic, or random in time, reflecting our ignorance about the detailed causes of individual mortality, reproduction, and dispersal. The description and prediction of stochastic population dynamics in time and space is fundamental to ecology and conservation biology. In this chapter we introduce the basic stochastic factors affecting population dynamics that will be analyzed in greater detail in the rest of the book. For example, as we will show, stochastic fluctuations can cause the extinction of a population that deterministically would persist indefinitely. This has important ramifications for conservation of small single populations and of metapopulations composed of multiple populations subject to local extinction and colonization. Stochasticity also can interact with human exploitation to cause the collapse or extinction of a population harvested under a strategy that deterministically would produce the maximum sustained yield. Failure to account for these basic features of stochastic population dynamics has been partially responsible for the declines of many threatened and endangered species and for the overexploitation and collapse of numerous living resources, including commercial fisheries.

Figure 1.1 displays several graphs of population fluctuations through time. For each species, the data represent annual counts or estimates of population size at a particular locality. Notice that populations of short-lived species such as the Great Tit and Blue Tit undergo primarily short-term, rapid fluctuations, whereas long-lived species such as the Mute Swan display a combination of short-term and long-term fluctuations. Tits are small passerine birds that begin reproduction at an age of 1 year and have a generation time, or average age at reproduction, of about 2 years. In contrast, the Mute Swan is a large bird that begins reproduction at an age of about 4 years and has a generation time of about a decade.

Populations and species differ substantially in their patterns of temporal fluctuation, whether they experience upward or downward trends, or fluctuate around a stable equilibrium. Variability in population size through time can be measured by the estimated coefficient of variation (CV), the sample standard deviation of population size among years divided by the sample mean population size, $\sigma_N/\bar{N}$. A similar measure of relative variability is the standard deviation of the natural logarithm of population size, $\sigma_{\ln N}$, which is nearly equivalent to the CV for small or moderate CV

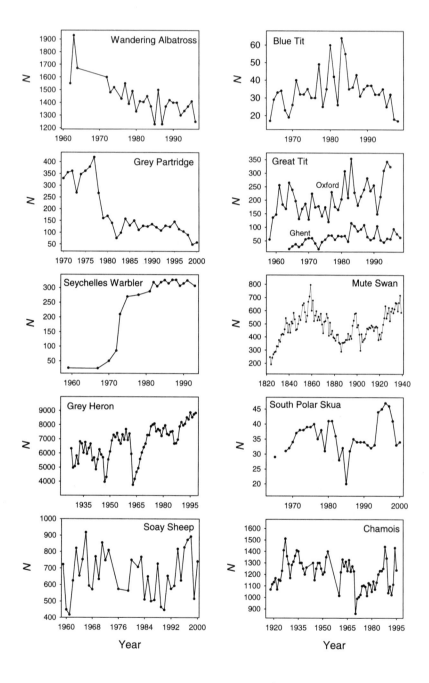

Year   Year

less than about 30%. CV in annual population estimates for a wide variety of animal species ranges from about 10% to 100% (Pimm and Redfearn, 1988; Pimm, 1991; Ariño and Pimm, 1995). Table 1.1 presents estimates of coefficients of variation of annual population censuses for bird species, each sampled over 20 years. The interpretation of such data is complicated because the estimated variance, and hence the estimated coefficient of variation, tends to increase with the timespan over which the data are collected (Pimm and Redfearn, 1988). This is a consequence of temporal autocorrelation in population size, meaning that population sizes in sequential years generally are not independent (see Chapter 3).

CV in bird species in Table 1.1 tends to decrease with increasing generation time, indicated by age at maturity, α, but large differences in CV exist even among species with the same α. (A linear regression of $\ln CV$ on α has significant slope -0.109 explaining 23% of the variance in $\ln CV$.) We will show that the major determinants of CV or $\sigma_{\ln N}$ are stochasticity in population growth rate and the strength of density-dependent population regulation (Eqs 1.9b, 2.2c, 4.10), both of which may be influenced by, or related to, life history (Sæther et al., 2002b). As already mentioned, life history differences among species strongly influence the pattern of population fluctuations described by the temporal autocorrelation of population size, which we introduce in this chapter and analyze in more detail in Chapter 3.

1.2 Concepts of stochasticity

The dynamics of every population has both deterministic (predictable) and stochastic (unpredictable) components that operate simultaneously. Populations below their carrying capacity, or average population size, tend to increase, as illustrated in Figure 1.1 by the Seychelles Warbler, which grew to a new level following conservation efforts (Komdeur, 1994). Populations above their carrying capacity, or with a negative intrinsic rate of increase, tend to decrease, as illustrated by the Wandering Albatross, which suffers increased adult mortality from eating bait hooked on long-line fishing

---

Fig. 1.1 Population fluctuations in eight bird species: the Wandering Albatross *Diomedea exulans* on South Georgia (Croxall *et al.*, 1997), Grey Partridge *Perdix perdix* in southern England (Aebischer and Potts, 1990; N. Aebischer, pers. com.), Seychelles Warbler *Acrocephalus sechellensis* on Cousin Island in the inner Seychelles (Komdeur, 1994), Grey Heron *Ardea cinera* in southern England (British Trust for Ornithology, pers. com.), Blue Tit *Parus caeruleus* in Ghent, Belgium (E. Matthysen, pers. com.), Great Tit *Parus major* at Ghent, Belgium (E. Matthysen, pers. com.) and in Wytham Wood at Oxford, England (Sæther *et al.*, 1998*a*; R. H. McCleery, pers. com.), Mute Swan *Cygnus olor* on the River Thames, England (Cramp, 1972), South Polar Skua *Catharacta maccormicki* on Pointe Géologie archipelago, Terre Adélie, Antarctica (H. Weimerskirch, pers. com.), and two mammals: Soay Sheep *Ovis aries* on Hirta Island, Scotland (T. Coulson, pers. com.), and Chamois *Rupicapra rupicapra* in the Swiss National Park, southern Switzerland (F. Filli, pers. com.).

Table 1.1 Coefficient of variation (CV) for fluctuations in annual population size in birds. CV is calculated only for the last 20 years of data. α = average age at first breeding of females

Species		α	Locality	Period	CV	Reference
Blue Tit	*Parus caeruleus*	1	Lower Saxony, Germany	1974–93	0.27	Winkel (1996)
Dipper	*Cinclus cinclus*	1	Lygnavassdraget, southern Norway	1978–97	0.46	Sæther et al. (2000b)
Garganey	*Anas querquedula*	1	Engure Marsh, Latvia	1974–93	0.48	Blums et al. (1993)
Great Tit	*Parus major*	1	Lower Saxony, Germany	1974–93	0.17	Winkel (1996)
Great Tit	*Parus major*	1	Wytham Wood, Oxford, England	1974–93	0.27	McCleery (pers. com.)
Northern Shoveler	*Anas clypeata*	1	Engure Marsh, Latvia	1974–93	0.38	Blums et al. (1993)
Nuthatch	*Sitta europaea*	1	Lower Saxony, Germany	1974–93	0.31	Winkel (1996)
Pied Flycatcher	*Ficedula hypoleuca*	1	Lower Saxony, Germany	1974–93	0.22	Winkel (1996)
Pied Flycatcher	*Ficedula hypoleuca*	1	Lingen, Germany	1974–93	0.25	Winkel (1998)
Pied Flycatcher	*Ficedula hypoleuca*	1	Kilpisjärvi, northern Finland	1968–87	0.45	Järvinen (1990)
Pochard	*Aythya ferina*	1	Engure Marsh, Latvia	1974–93	0.51	Blums et al. (1993)
Seychelles Warbler	*Acrocephalus sechellensis*	1	Cousin Island, Seychelles	1973–93[1]	0.10	Komdeur (1994)
Song Sparrow	*Melospiza melodia*	1	Mandarte Island, British Columbia, Canada	1979–98	0.49	Sæther et al. (2000a)
Tufted Duck	*Aythya fuligula*	1	Engure Marsh, Latvia	1974–93	0.25	Blums et al. (1993)
Avocet	*Recurvirostra avosetta*	2	Havergate, Suffolk, England	1967–86	0.30	Hill (1988)
Grey Heron	*Ardea cinera*	2	Southern England	1979–98	0.09	B.T.O. (pers. com.)
Kentish Plover	*Charadrius alexandrinus*	2	Niedersachsen, Germany	1974–93	0.57	Hälterlein and Südbeck (1996)
Ural Owl	*Strix uralensis*	2	Hämeenlinna, Finland	1969–88	0.50	Saurola (1989)
Common Tern	*Sterna hirundo*	3	Mecklenburg-Vorpommern, Germany	1978–97	0.39	Spretke (1998)
Sandwich Tern	*Sterna sandvicensis*	3	Schleswig-Holstein, Germany	1974–93	0.44	Hälterlein and Südbeck (1996)
Mute Swan	*Cygnus olor*	4	River Thames, England	1920–39	0.15	Cramp (1972)
South Polar Skua	*Catharacta maccormicki*	6	Pointe Géologie, Terre Adélie, Antarctica	1981–2000	0.19	Weimerskirch (pers. com.)
Short-tailed Shearwater	*Puffinus tenuirostris*	7	Fisher Island, Tasmania	1965–84	0.15	Bradley et al. (1991)
Northern Fulmar	*Fulmarus glacialis*	9	Eynhallow, Orkney	1958–77	0.27	Dunnet et al. (1979)
Wandering Albatross	*Diomedea exulans*	10	Bird Island, South Georgia	1974–93	0.06	Croxall et al. (1997)

[1] Only 15 censuses available during the last 20 study years.

gear (Croxall *et al.*, 1997), and the Barn Swallow (Fig. 5.1), for which insect prey are reduced by modern agriculture (Møller, 2001). Such trends reflect the deterministic component of population dynamics.

Populations also may be subject to deterministic cycles. For example, many insect species with generation times less than 1 year, undergo cyclic fluctuations driven by seasonal environmental factors such as temperature, precipitation, and food availability (Myers, 1998). We largely avoid the complication of environment-ally driven, seasonal cycles by focusing in this book mainly on vertebrate species with synchronized annual reproduction. Specialized predator–prey interactions in simple ecosystems can create sustained population cycles such as for the Canadian Lynx (*Lynx canadensis*) and Snowshoe Hare (*Lepus americanus*) (reviewed by Case, 2000). Simple predator–prey cycles do not commonly occur in diverse ecosystems where predators have multiple prey species and prey have multiple competitors. We show, however, in Chapter 3 that internally driven population cycles can be produced by time lags inherent in the life history of a species, and by strong density-dependent population regulation.

In a constant environment, deterministic dynamics of populations without age struc-ture can produce sustained fluctuations, particularly for populations with potentially high growth rates reproducing at discrete time intervals. May (1976, 1981; May and Oster, 1976) analyzed a variety of simple discrete-time population models, showing that as the intrinsic growth rate increases, from much less than 1 per unit time to more than 1 or 2 per unit time, the dynamics change from smooth asymptotic approach to a stable equilibrium, through regular cycles of increasing length, with period 2, 4, 8, . . . time units, finally becoming chaotic. Chaos has much the same appearance as unpre-dictable, stochastic dynamics, but is generated by deterministic mechanisms involving large overshoots and undershoots of an unstable carrying capacity. In recent decades chaos has received a great deal of attention from population theorists (reviewed by Ellner and Turchin, 1995; Bjørnstad and Grenfell, 2001; Dennis *et al.*, 2001). How-ever, analysis of insect populations with potentially high growth rates indicates that the vast majority of wild populations are not cyclic or chaotic (Hassell *et al.*, 1976). Similar results were obtained by Ellner and Turchin (1995) even allowing for a com-bination of cycles or chaos and stochasticity. The primary reason for this is that wild populations, even those with extremely high fecundity, usually experience high density-independent mortality which severely limits the maximum rate of population growth (Myers *et al.*, 1999). Vertebrate species with synchronized annual repro-duction and adult body weights above one kg also tend to have maximum rates of population growth less than 0.1 per year (Charnov, 1993), implying that they usually are not subject to intrinsically driven deterministic cycles or chaos. Finally, it has been shown theoretically that in discrete-time population models with chaotic dynamics, extinction is most likely to occur from large population size in a single time step, fol-lowing a large overshoot of carrying capacity due to a high rate of population growth (Ripa and Heino, 1999). In contrast, as illustrated in Figure 2.3, most documented extinctions of both vertebrates and invertebrates occur from a small population size after a gradual decline over an extended time. We can therefore be confident that most apparently stochastic population fluctuations are actually stochastic and not either noisy cycles or chaos.

Data on population dynamics include three basic forms of stochasticity: demographic stochasticity, environmental stochasticity, and sampling error (May, 1973, 1974; Turelli, 1977; Leigh, 1981; Shaffer, 1981).

(1) Demographic stochasticity refers to chance events of individual mortality and reproduction, which are usually conceived as being independent among individuals. In any given time period, an individual either dies or survives with a certain probability. Individuals also have a probability distribution of number of offspring produced per unit time. Because demographic stochasticity operates independently among individuals, it tends to average out in large populations and has a greater impact on small populations.

(2) Environmental stochasticity refers to temporal fluctuations in the probability of mortality and the reproductive rate of all individuals in a population in the same or similar fashion. The impact of environmental stochasticity is roughly the same for small and large populations. It therefore constitutes an important risk of population decline in all populations regardless of their abundance at a given location. Wide geographic range can reduce the risk of population decline if environmental stochasticity is weakly correlated among spatial locations. Spatial autocorrelation in environmental stochasticity is therefore a major determinant of spatial autocorrelation in population size.

Unpredictable catastrophes that suddenly reduce a population by a substantial fraction, such as might occur in a drought, fire, flood, landslide, volcano, hurricane, or epidemic, are sometimes classified as a separate form of stochasticity (Shaffer, 1981; Mangel and Tier, 1993; Young, 1994; Ludwig, 1996a). Here we usually include random catastrophes as an extreme form of environmental stochasticity (Shaffer, 1987), justifying this approach in Chapter 2.

(3) Measurement error in estimates of population size or density, based on a sampling procedure, such as mark–recapture methods, or sampling a proportion of the potentially suitable habitat for a species (Seber, 1982). For populations in which a complete and accurate census is available, such as for most of the data considered in this book, measurement error can be ignored. This permits us to avoid complicated statistical procedures for dealing with measurement error in population time series, such as state–space analysis (Box et al., 1994) and Markov chain Monte Carlo methods (Gilks et al., 1996), facilitating our focus on basic biological models. We show, however, in Chapter 6 that uncertainty in population estimates substantially influences sustainable harvesting strategies.

1.3 Formulation of stochasticity

Consider initially a population with synchronized annual reproduction and no age structure. Two distinct situations exist in which age structure can be ignored. The first is nonoverlapping generations where individuals of age 1 year reproduce and then die, as for univoltine insects or annual plants with no seed bank. The second is overlapping generations where individuals mature in a single year but thereafter may survive multiple years, with survival and reproductive rates independent of age, as closely approached by many small or medium-sized vertebrates, including passerine birds such as the Great Tit, Blue Tit, and Song Sparrow, in Figure 1.1. In such species, age effects are of little demographic consequence (Sæther, 1990).

1.3.1 Stochasticity in adult female population size

For a population with synchronized annual reproduction and no age structure, we assume that the adult female population is censused during the breeding season immediately following reproduction. Population size changes because of the net effects of individual reproduction and survival, which are the components of fitness in a closed population. Individual fitness per year can be defined as individual viability to the next breeding season (1 for surviving and 0 for dead) plus the number of female offspring that survive their first year. As with most demographic models, we follow only females and their female offspring, assuming that males are not limiting in reproduction, i.e. that all females have access to mates. Individual fitness in a given year can be statistically decomposed into an expected value plus an individual deviation. The fitness of individual i in a given year is

$$w_i = \mu_w + \delta_i \qquad\qquad 1.1$$

where μ_w is the expected individual fitness and the deviation from the expected fitness for the ith individual is δ_i which has an expected value of zero, $E[\delta_i] = 0$. The expected fitness μ_w and the individual deviations δ_i are independent random variables. Random variation in the expected fitness that is independent of population density constitutes environmental stochasticity. Random variation in individual fitness, coupled with sampling effects in a finite population, produces demographic stochasticity.

Figure 1.2 presents examples of the distribution of individual female fitness per year in populations of two species, the Song Sparrow and the Great Tit, based on data taken during sequential breeding seasons for individual birds identified by colored leg rings. Both of these species reach maturity and begin reproduction at age 1 year, so that variation among individuals due to age is minimal. The distribution of individual fitness undergoes yearly changes not only in the mean value within years, but also in shape and particularly in variance. In a given year the distribution of individual fitness often deviates strongly from normality, and does not follow any simple type of distribution, frequently having a substantial proportion of individuals with zero fitness. The same generally applies to the distribution of individual lifetime fitness (Clutton-Brock, 1988; Newton, 1989). For the years illustrated in Figure 1.2, changes in the mean fitness among years are caused by a combination of changing population density and environmental stochasticity.

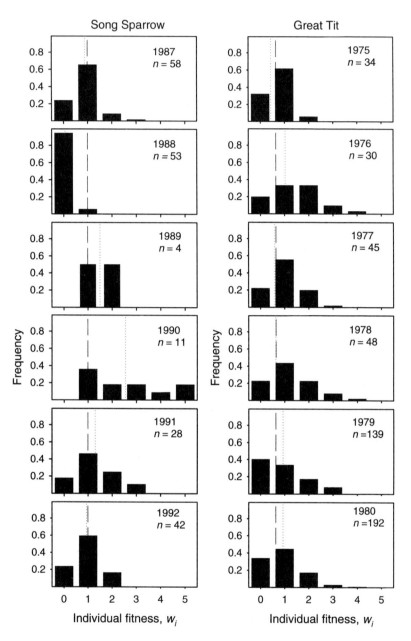

Fig. 1.2 Annual variation in the distribution of individual fitness of two passerine species, the Song Sparrow *Melospiza melodia* on Mandarte island and the Great Tit in Wytham Wood. The dashed line indicates the mean fitness across all years and the dotted line the mean fitness of single years. The number of females is *n*.

The finite rate of population increase, or multiplicative growth rate, of a population of size N in a given year is the mean individual fitness

$$\lambda = \bar{w} = \frac{1}{N} \sum_{i=1}^{N} w_i = \mu_w + \frac{1}{N} \sum_{i=1}^{N} \delta_i \qquad 1.2$$

so the population at a given time t grows according to $N(t+1) = \lambda(t)N(t)$. This and similar dynamic equations for population size are valid if N either enumerates females or represents total population size with a nearly constant sex ratio. The expected multiplicative growth rate of the population averaged over the probability distribution of environments among years is $\bar{\lambda} = E[\mu_w]$. In terms of environmental effects on population growth, $\bar{\lambda}$ also can be called the multiplicative growth rate in the average environment, but this terminology does not apply to any other measure of the environment such as physical factors like rainfall, as explained after equations 1.5.

Stochasticity in population dynamics arises from yearly variation in the multiplicative growth rate that is not related to density-dependent changes in fitness. This is caused by environmental fluctuations affecting the mean individual fitness, described by the environmental variance among years, $\sigma_e^2 = \mathrm{Var}[\mu_w]$, and by sampling effects within years in a finite population due to the demographic variance defined as the expected variance in individual fitness within years, $\sigma_d^2 = \mathrm{Var}[\delta_i]$. Taking variances on both sides of equation 1.2, using the independence of μ_w and δ_i, gives

$$\sigma_\lambda^2 = \mathrm{Var}[\mu_w] + \frac{1}{N^2} \sum_{i=1}^{N} \mathrm{Var}[\delta_i] = \sigma_e^2 + \frac{\sigma_d^2}{N} \qquad 1.3$$

(May, 1973, 1974; Turelli, 1977; Leigh, 1981; Goodman, 1987). In the absence of environmental stochasticity equation 1.3 is the standard formula for the variance of the mean (fitness) in a random sample of N individuals from a hypothetical infinite population with (demographic) variance σ_d^2 (Stuart and Ord, 1994).

Separating density dependence from demographic and environmental stochasticity as causes of differences among years in the distributions of individual fitness is more difficult, requiring joint statistical estimation of deterministic and stochastic parameters in population dynamics. Statistical methods for estimating population parameters using data on individual fitness within years, and time series of population censuses across years, are addressed at the end of this chapter, after a brief introduction to density-dependent population dynamics in a random environment. Estimates in Table 1.2 obtained by these methods indicate that the demographic variance averaged over a span of years usually is in the range of 0.1–1, whereas the environmental variance often is an order of magnitude smaller.

Equation 1.3 shows that environmental stochasticity strongly affects λ in all populations regardless of size, whereas demographic stochasticity most strongly affects λ in small populations. In sufficiently large populations, with $N \gg \sigma_d^2/\sigma_e^2$, environmental stochasticity is much more important than demographic stochasticity. This allows us to define a critical population size $N_c = 10\sigma_d^2/\sigma_e^2$ above which

Table 1.2 Estimated demographic variance, $\hat{\sigma}_d^2$, and environmental variance, $\hat{\sigma}_e^2$, in multiplicative growth rate in populations with different mean age at first breeding of females, α

Species	Locality	α	$\hat{\sigma}_d^2$	$\hat{\sigma}_e^2$	Reference
Barn Swallow, *Hirundo rustica*	Denmark	1	0.18	0.024	Engen *et al.* (2001)
Dipper	Southern Norway	1	0.27	0.21	Sæther *et al.* (2000*b*)
Great Tit	Wytham Wood, U.K.	1	0.57	0.079	Sæther *et al.* (1998*a*)
Pied Flycatcher	Hoge Veluwe, Netherlands	1	0.33	0.036	Sæther *et al.* (2002*a*)
Song Sparrow	Mandarte Island, B.C.	1	0.66	0.41	Sæther *et al.* (2000*a*)
Soay Sheep, *Ovis aries*	Hirta Island, U.K.	1	0.28	0.045	Sæther *et al.* (unpubl.)
Brown Bear, *Ursus arctos*	Southern Sweden	4	0.16	0.003	Sæther *et al.* (1998*b*)
Brown Bear	Northern Sweden	5	0.18	0.000	Sæther *et al.* (1998*b*)

demographic stochasticity can be neglected. Estimates of these parameters in Tables 1.2 and 1.4 indicate that N_c is often on the order of 100, but may range from about 10 to 1000 in populations of different species.

1.3.2 Demographic stochasticity with two sexes

If the sexes are not distinguished, the demographic stochasticity in multiplicative growth rate for total adult population size includes not only the variance in individual fitness among adult females but also involves the adult sex ratio. With overlapping generations the adult sex ratio is not independent among years. To avoid this complication we consider only the case of nonoverlapping generations where mature individuals die after reproduction at age 1. For simplicity we also assume a polygamous mating system in which female reproduction is not limited by access to mates. Redefining the fitness of the ith adult female, $w_i = d_i + s_i$, as the number of her daughters, d_i, and sons, s_i, that survive to age 1, the fitness of an adult male then must be defined as 0. The expected fitness of adults, or multiplicative growth rate of the total adult population in a given environment is therefore

$$\lambda = p\bar{w} + (1 - p)0 = p\bar{w}$$

where p is the expected frequency of females in the adult population. The total demographic variance, denoted as σ_D^2, is the variance of individual fitness for adults of both sexes in a given environment (excluding environmental stochasticity),

$$\sigma_D^2 = p\mathrm{E}[w_i^2] + (1 - p)0^2 - \lambda^2$$
$$= p(\sigma_w^2 + \bar{w}^2) - \lambda^2$$
$$= p\sigma_w^2 + \frac{1 - p}{p}\lambda^2$$

where we have substituted $\bar{w} = \lambda/p$. The variance of female fitness in a given environment can be expressed as $\sigma_w^2 = \sigma_d^2 + \sigma_s^2 + 2\text{Cov}[d_i, s_i]$ where σ_d^2 is the standard demographic variance accounting only for females, σ_s^2 is the variance in number of sons per adult female, and $\text{Cov}[d_i, s_i]$ is the covariance in number of daughters and sons per adult female.

It is informative to consider the simplest case of a stable (or constant) population with a Mendelian mechanism of sex determination and equal viability of male and female offspring, so that the mean female fitness is $\bar{w} = 2$, with $p = 1/2$ and $\lambda = 1$. Further assume that female fitness has a Poisson distribution, with the probability of w progeny being $e^{-2}2^w/w!$ so the variance of w equals its mean, $\sigma_w^2 = 2$. These assumptions imply that $\sigma_s^2 = \sigma_d^2 = 1$ and $\text{Cov}[d_i, s_i] = 0$. Then the standard demographic variance accounting only for females is unity, $\sigma_d^2 = 1$, whereas the demographic variance of the total population is twice as large, $\sigma_D^2 = 2$. This illustrates that the demographic variance of the total population, including both sexes, typically is substantially greater than that for females only.

1.4 Stochasticity on the natural logarithmic scale

Population size often is measured on the natural logarithmic scale, $X = \ln N$, on which the dynamics are simply $X(t + 1) = X(t) + r(t)$ where $r(t) = \ln \lambda(t)$ is the population growth rate on the log scale. Stochasticity in population dynamics on the log scale can be characterized in the following way. The multiplicative growth rate in a given year can be expressed as the mean, $\bar{\lambda}$, plus an annual deviation from the mean, $\varepsilon(t)$,

$$\lambda(t) = \bar{\lambda} + \varepsilon(t)$$

where the annual deviation, $\varepsilon(t)$, has mean 0 and variance given by equation 1.3. The population growth rate on the log scale can be rewritten and expanded in a Taylor series as

$$r(t) = \ln \bar{\lambda} + \ln[1 + \varepsilon(t)/\bar{\lambda}]$$
$$= \ln \bar{\lambda} + \varepsilon(t)/\bar{\lambda} - [\varepsilon(t)/\bar{\lambda}]^2/2 + \cdots$$

The mean and variance of population growth rate on the log scale, $\bar{r} = \text{E}[\ln \lambda(t)]$ and $\sigma_r^2 = \text{Var}[\ln \lambda(t)]$, are then approximately

$$\bar{r} \cong \ln \bar{\lambda} - \frac{\sigma_r^2}{2} \quad \text{and} \quad \sigma_r^2 \cong \frac{\sigma_\lambda^2}{\bar{\lambda}^2}. \qquad 1.4$$

Thus demographic and environmental stochasticity reduce the mean growth rate of a population on the logarithmic scale, compared with that in the (constant) average environment. This implies that stochasticity tends to reduce most population trajectories and therefore constitutes an important risk of extinction, as will be elaborated below and in Chapters 2 and 3.

These approximations for the mean and variance of population growth rate on the log scale are accurate when the fluctuations in multiplicative growth rate are not very large relative to the expected multiplicative growth rate. Their accuracy can be checked by using a particular probability distribution for multiplicative growth rate, which is always non-negative. For example, if $\lambda(t)$ has a lognormal distribution the exact formulas for the mean and variance of population growth rate on the log scale are (Johnson and Kotz, 1970; Lande and Orzack, 1988)

$$\bar{r} = \ln \bar{\lambda} - \frac{\sigma_r^2}{2} \quad \text{and} \quad \sigma_r^2 = \ln \left(1 + \frac{\sigma_\lambda^2}{\bar{\lambda}^2} \right). \qquad 1.5a$$

Alternatively, if $\lambda(t)$ randomly takes only two possible values, $\bar{\lambda} + \varepsilon$ and $\bar{\lambda} - \varepsilon$, each with probability 1/2 (with $0 \le \varepsilon \le \bar{\lambda}$), then $\sigma_\lambda^2 = \varepsilon^2$ and the mean and variance of population growth rate on the log scale are

$$\bar{r} = \ln \bar{\lambda} + \frac{1}{2} \ln \left(1 - \frac{\sigma_\lambda^2}{\bar{\lambda}^2} \right) \quad \text{and} \quad \sigma_r^2 = \frac{1}{4} \left[\ln \left(\frac{\bar{\lambda} + \sigma_\lambda}{\bar{\lambda} - \sigma_\lambda} \right) \right]^2. \qquad 1.5b$$

A little numerical investigation demonstrates that the approximations in equations 1.4 have good accuracy when the CV of multiplicative growth rate, $\sigma_\lambda/\bar{\lambda}$, is less than about 30%, as shown in Table 1.3.

The details of how stochasticity is introduced into a deterministic model can have major consequences for the dynamics. If instead of adding environmental noise to the multiplicative growth rate (as above), the noise is directly added to the population growth rate on the log scale, $r(t) = \bar{r} + \varepsilon(t)$, with the mean environmental effect equal to zero, $\bar{\varepsilon} = 0$, then it would be found that environmental stochasticity increases the mean multiplicative growth rate, $\bar{\lambda}$ (Higgins *et al.*, 2000; Efford, 2001). For example, assuming $\varepsilon(t)$ has a normal distribution with mean $\bar{\varepsilon}$ and variance σ_ε^2 using the first of equations 1.5a gives $\bar{\lambda} = e^{\bar{r}+\bar{\varepsilon}+\sigma_\varepsilon^2/2}$. To ensure that changing the environmental variance does not alter $\bar{\lambda}$ it is therefore necessary to set $\bar{\varepsilon} = -\sigma_\varepsilon^2/2$ instead of $\bar{\varepsilon} = 0$. Consider the Ricker equation often employed in the fisheries literature to predict juvenile recruits, $R(t + 1)$, from the adult stock, $N(t)$, according

Table 1.3 Mean and variance of population growth rate $r(t)$ from the exact formulas (eqs 1.5a,b), compared with the general approximation (eqs 1.4) for different values of the coefficient of variation of $\lambda(t)$

	Mean, $\bar{r} - \ln \bar{\lambda}$				Variance, σ_r^2			
$\sigma_\lambda/\bar{\lambda} =$	0.1	0.3	0.5	1.0	0.1	0.3	0.5	1.0
Lognormal	−0.00498	−0.0431	−0.112	−0.347	0.00995	0.0862	0.223	0.693
Two state	−0.00503	−0.0472	−0.144	−∞	0.01007	0.0958	0.302	∞
Approximate	−0.005	−0.045	−0.125	−0.5	0.01	0.09	0.25	1

to $R(t+1) = N(t)e^{a-bN(t)}$. At any given stock size, adding random noise to the parameter a would increase the expected number of recruits, whereas adding noise to $A = e^a$ would not alter the expected number of recruits.

Environmental stochasticity is commonly modeled using an explicit environmental variable. In models where population growth rate is a nonlinear function of an environmental variable, the mean values of both r and λ generally will differ from their values in the average environment with respect to the environmental variable. For example, Caughley (1987) described population growth rate r for Red Kangaroos (*Macropus rufus*) as a concave (upwards) function of rainfall, v, through its effect on pasture. Variability in rainfall then necessarily decreases $\bar{r}$ compared with the population growth rate at the average rainfall (Davis *et al.*, 2002), as can be seen for small fluctuations in rainfall by expanding $r(v)$ in a Taylor series around $\bar{v}$ and taking expectations, $\bar{r} \cong r(\bar{v}) + \frac{1}{2}\sigma_v^2(\partial^2 r/\partial v^2)_{\bar{v}}$, so that if the second derivative is negative then $\bar{r} < r(\bar{v})$.

Thus in assessing the influence of environmental stochasticity on population growth rate it is necessary to define clearly the measure of population growth rate, how it is affected by environmental stochasticity, and whether the average environment should be interpreted with respect to environmental effects on population growth rate or an explicit environmental variable.

1.5 Density-independent growth in a random environment

Now consider the dynamics of the basic model of density-independent population growth in a random environment, $N(t+1) = \lambda(t)N(t)$ (Lewontin and Cohen, 1969). The multiplicative growth rate, $\lambda(t)$, is assumed to have the same probability distribution each year due to environmental stochasticity, with no temporal autocorrelation in the environment, and we suppose that the population size is large enough to neglect demographic stochasticity. Figure 1.3 illustrates five simulated population trajectories, or sample paths, when $\lambda(t)$ has a lognormal distribution. Population trajectories are depicted on a log scale on which the trajectory for the corresponding deterministic model is a straight line, with a multiplicative growth rate equal to that in the average environment, $\bar{\lambda}$, with $\sigma_\lambda^2 = 0$. After sufficient time all five sample paths are below the trajectory for the deterministic model. This illustrates that environmental stochasticity tends to reduce the population size in the long run.

The dynamics of this model are most conveniently analyzed on the natural log scale, $X(t+1) = X(t) + r(t)$. In the absence of demographic stochasticity, and with a stationary distribution of environmental effects, the density-independent growth rate, $r(t)$, has a constant mean, $\bar{r}$, and a constant environmental variance, σ_r^2. For this reason population dynamics are often analyzed on a logarithmic scale (Royama, 1992). Starting from an initial log population size $X(0)$ the population trajectories are

$$X(t) = X(0) + \sum_{\tau=0}^{t-1} r(\tau). \qquad \text{1.6a}$$

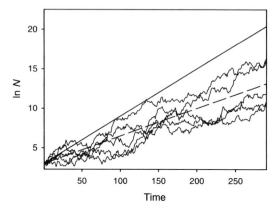

Fig. 1.3 Simulation of five population trajectories describing density independent growth in a random environment, neglecting demographic stochasticity, $N(t + 1) = \lambda(t)N(t)$, with parameters $\bar{\lambda} = 1.06, \sigma_e^2 = 0.05, N_0 = 20$. Solid straight line shows the deterministic trajectory for population growth in a constant average environment ($\sigma_e^2 = 0$); dashed line shows the expected long-run growth.

Taking expectations of both sides, and separately taking variances of both sizes employing the assumption of no environmental autocorrelation, produces

$$\bar{X}(t) = X(0) + \bar{r}t \quad \text{and} \quad \sigma_X^2(t) = \sigma_r^2 t. \qquad 1.6b$$

Both the mean and variance of log population size increase linearly with time. If environmental stochasticity in population growth rate $r(t)$ is normally distributed, then the log population size is exactly normally distributed at all times, because a sum of independent, normally distributed random variables also has a normal distribution (Stuart and Ord, 1994). Figure 1.4(A) depicts for this situation how the probability distribution of log population size changes through time. For any distribution of environmental stochasticity, after several time steps the distribution of $X(t)$ approaches normality. This follows from the Central Limit Theorem of statistics, according to which a sum of independent, identically distributed random variables is asymptotically normal (Stuart and Ord, 1994).

Considering two different scales for describing population size, N and X, related by a monotonic transformation, preservation of probability density in the corresponding infinitesimal intervals of each variable requires that $\varphi(X)dX = f(N)dN$. For example if $X = \ln N$ has a normal distribution $\varphi(X)$ then $dX/dN = 1/N$ and population size has the lognormal distribution $f(N) = \varphi(\ln N)/N$.

1.5.1 Long-run growth rate

The slope of a stochastic population trajectory between the initial and final points at time t on the log scale is $[X(t) - X(0)]/t$. From equations 1.6b this slope has

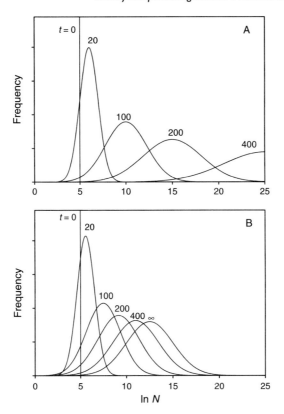

Fig. 1.4 (A) Distribution of log population size starting from the initial log size $\ln N_0 = 5$ under density-independent growth in a random environment (eqs 1.6) after $t = 0, 20, 100, 200,$ and 400 years. Parameters are $\bar{r} = 0.05$ and environmental variance $\sigma_r^2 = 0.05$ with no demographic stochasticity. (B) Distribution of log population size under density regulation of the Gompertz type (eqs 1.9) after $t = 0, 20, 100, 200,$ and 400 years, and the stationary distribution after ∞ (infinity) years. Parameters are identical to those in panel A, but with density dependence, $\gamma = 0.004.$

expectation $\bar{r}$ and variance σ_r^2/t. Thus with increasing time the variance of the slope approaches zero and the slope of all population trajectories approaches $\bar{r}$. For this reason $\bar{r}$ is often called the long-run growth rate of a population (Cohen, 1977, 1979; Tuljapurkar, 1982). When $\bar{r}$ is positive all population trajectories eventually increase and when $\bar{r}$ is negative all population trajectories eventually decline toward extinction. The expected population growth rate on the log scale, $\bar{r}$, therefore accurately characterizes the tendency for a population to increase or decrease in a stochastic environment.

This result can be compared with the dynamics of the expected population size. Taking expectations of the recursion on the original, untransformed scale, using independence of $\lambda(t)$ and $N(t)$, gives $\bar{N}(t+1) = \bar{\lambda}\bar{N}(t)$, which has the solution

$$\bar{N}(t) = \bar{\lambda}^t N(0). \qquad 1.7$$

Despite environmental stochasticity the expected population size grows exactly as in the deterministic model in the average environment.

A fundamental conclusion from equations 1.6b and 1.7 is that the dynamics of the expected population size in the stochastic model can be extremely misleading (see Fig. 1.3). This occurs because environmental stochasticity generally causes the long-run growth rate, $\bar{r}$, to be less than $\ln \bar{\lambda}$ (see eq. 1.4). Lewontin and Cohen (1969) emphasized the counter-intuitive result that in this basic model the expected population size can increase toward infinity even though all population trajectories eventually decline toward extinction. This would happen when the average environment is conducive to population growth, $\bar{\lambda} > 1$, but there is sufficient environmental stochasticity that the long-run growth rate is negative, $\bar{r} < 0$. Such results emphasize the importance of stochastic models for understanding population dynamics. In Chapter 3 the concept of the long-run growth rate is extended to age-structured populations.

1.5.2 Autocorrelated environment

An essential distinction exists between environmental autocorrelation and population autocorrelation, which are sometimes confused in the literature. Environmental autocorrelation refers to the temporal pattern of environmental stochasticity driving population dynamics; this may be caused by autocorrelation in physical and/or biological factors, primarily weather and ecological interactions with other species. Population autocorrelation refers to the temporal pattern of population sizes. Population autocorrelation may be caused by environmental autocorrelation but can also occur in completely random, uncorrelated environments because of density dependence (see Section 1.6) and life history (see Chapter 3). After removing seasonal cycles, most terrestrial localities on continents show little autocorrelation in weather patterns over times longer than 1 year (Namias, 1978; Nicholls, 1980). Physical factors are therefore not likely to create much population autocorrelation for species with synchronized annual reproduction. The complexity of ecological interactions for most vertebrate species, in comparison for example with insect species specialized to a single host plant, tends to reduce the possible environmental autocorrelation due to biological factors. It is therefore not surprising that our analysis of population time series in Chapter 3 finds that in terrestrial vertebrate species most population autocorrelation is caused by density dependence and time delays in life history, with little evidence of environmental autocorrelation.

Environmental autocorrelation can be readily incorporated in the basic density-independent model. The environmental autocorrelation between population growth rates $r(t)$ and $r(t - \tau)$ separated by a time lag τ is denoted as $\rho_e(\tau)$. For a stationary distribution of environments the autocorrelation is generally symmetric in the time lag, $\rho_e(-\tau) = \rho_e(\tau)$ with $\rho_e(0) = 1$, and in the absence of deterministic trends or cycles the autocorrelation generally approaches zero for long lags. Taking expectations in equation 1.6a yields the same expected dynamics as in equation 1.6b. Upon taking variances, however, extra terms appear for the covariances between pairs of growth rates at different times. It is not difficult to show that for sufficiently long times, greater than the time scale of environmental autocorrelation, the variance still increases linearly with time but at a magnified rate,

$$\bar{X}(t) = X(0) + \bar{r}t \quad \text{and} \quad \sigma_{\bar{X}}^2(t) \approx \left[1 + 2\sum_{\tau=1}^{\infty} \rho_e(\tau)\right]\sigma_r^2 t. \qquad 1.8$$

Using the same argument as in the case of no environmental autocorrelation, over a timespan t the slope of stochastic population trajectories, $[X(t) - X(0)]/t$, has expectation $\bar{r}$ and variance $\left[1 + 2\sum_{\tau=1}^{\infty}\rho_e(\tau)\right]\sigma_r^2/t$ so that as the timespan increases the variance of the slope approaches 0 and the slope of all trajectories approaches $\bar{r}$. Thus, for populations without age structure, environmental autocorrelation does not alter the long-run growth rate.

Again $X(t)$ is asymptotically normal by the Central Limit Theorem, so $\bar{X}(t)$ is the asymptotic median log population size corresponding to the median population size $e^{\bar{X}(t)}$, which is not altered by environmental autocorrelation. However, environmental autocorrelation does alter $\sigma_{\bar{X}}^2(t)$. This implies that environmental autocorrelation affects both the mean and variance of population size. To see this, suppose for simplicity that successive values of $r(t)$ have a multivariate normal distribution so that $X(t)$ is normally distributed at all times (Stuart and Ord, 1994). Then using transformations analogous to equations 1.5 the mean and variance of population size are

$$\bar{N}(t) = e^{\bar{X}(t)+(1/2)\sigma_{\bar{X}}^2(t)} \quad \text{and} \quad \sigma_N^2(t) = [\bar{N}(t)]^2[e^{\sigma_{\bar{X}}^2(t)} - 1].$$

Thus positive environmental autocorrelation, which increases $\sigma_{\bar{X}}^2(t)$, also increases both the mean and variance of population size, while negative environmental autocorrelation has the opposite effect.

The approximation in equations 1.8 substantially broadens the applicability of models without age structure in temporally uncorrelated (serially independent) environments that will be extensively discussed in subsequent chapters. Models incorporating this approximation should be reasonably accurate when the time scale of environmental autocorrelation is much shorter than that for extinction or density dependence (Turelli, 1977; Lande and Orzack, 1988; Foley, 1994, 1997).

1.6 Density-dependent growth in a random environment

Populations with a positive long-run growth rate ultimately become limited by density-dependent competition among individuals for resources such as food and space. An especially simple model is obtained by adding to the stochastic growth rate a deterministic limitation of population growth rate with the Gompertz form, that is, linear on the log scale, $X(t + 1) = (1 - \gamma)X(t) + r(t)$, where $X(t) = \ln N(t)$ supposing that $X(t) \geq 0$. Here $r(t)$ is the density-independent growth rate, also called the intrinsic rate of increase or intrinsic growth rate, and γ designates the strength of density dependence, with $0 < \gamma < 2$ (Royama, 1992). Assuming the environmental distribution is stationary, with no autocorrelation, and the population size remains large enough to neglect demographic stochasticity, so that the density-independent growth rate $r(t)$ has a constant mean and variance, $\bar{r}$ and σ_r^2, recursions for the mean and variance of log population size are

$$\bar{X}(t + 1) = (1 - \gamma)\bar{X}(t) + \bar{r} \quad \text{and} \quad \sigma_X^2(t + 1) = (1 - \gamma)^2\sigma_X^2(t) + \sigma_r^2. \qquad 1.9a$$

With an initial log population size $X(0) = X_0$ these have the solutions

$$\bar{X}(t) = \frac{\bar{r}}{\gamma} + \left[X_0 - \frac{\bar{r}}{\gamma}\right](1 - \gamma)^t \quad \text{and} \quad \sigma_X^2(t) = \frac{\sigma_r^2}{2\gamma - \gamma^2}[1 - (1 - \gamma)^{2t}].$$
$$1.9b$$

On a time scale of $1/\gamma$ years, the log population size approaches a stationary distribution with mean $\bar{r}/\gamma$ and variance $\sigma_r^2/(2\gamma - \gamma^2)$. Furthermore, if the distribution of $r(t)$ is normal then the distribution of $X(t)$ also is normal (see Fig. 1.4B).

The autocorrelation between log population sizes $X(t)$ and $X(t - \tau)$ separated by a time lag τ is denoted as $\rho(\tau)$. With no environmental autocorrelation in the intrinsic growth rate $r(t)$, it can be shown from the basic dynamics (above eq. 1.9a) that the population autocorrelation function, which must be symmetric with $\rho(\tau) = \rho(-\tau)$ and $\rho(0) = 1$, obeys the recursion $\rho(\tau + 1) = (1 - \gamma)\rho(\tau)$ with the solution $\rho(\tau) = (1 - \gamma)^{|\tau|}$. Thus density dependence creates population autocorrelation even in the absence of environmental autocorrelation and age structure. This population autocorrelation function approaches zero on a time scale of about $1/\gamma$ years. Weak density dependence, $\gamma \ll 1$, entails a slow expected rate of return to the mean log population size and a tendency for population trajectories to undergo long excursions away from the equilibrium. We examine population autocorrelation in more detail for density-dependent age-structured populations in Chapter 3.

1.7 Parameter estimation

1.7.1 Demographic and environmental stochasticity

The estimation of demographic and environmental variances is difficult for at least four reasons. First, reliable estimates of the demographic variance require observations on individual survival and reproduction, which may not be feasible to obtain for many species especially when juveniles disperse from their natal areas. Second, information about the environmental variance must be extracted from observations of annual changes in population size. Time series that are considered long by ecologists, are actually rather short from a statistical point of view. Consequently, the environmental variance cannot be estimated with high precision. The third problem is that exact time series of population fluctuations only can be recorded for a small number of species. A complete census usually cannot be performed, which entails that only time series of rather uncertain population estimates are available. When the sampling variance is large compared with the environmental variance, estimation of the environmental variance will require large data sets. For these reasons we analyze data from species of birds and mammals for which individual survival and reproduction as well as complete population censuses are available. Finally, it is necessary to separate density-dependent changes in population size from those caused by demographic and environmental stochasticity. Even if environmental variance is the main source of stochasticity during the period of data collection, the demographic variance will still dominate as the population approaches extinction. Unfortunately, omission of the demographic variance would then lead to serious underestimation of the probability of extinction and overestimation of the expected time to extinction (see Chapters 2 and 5, and Engen *et al.*, 2001).

The demographic variance reflects differences among individuals in reproduction and survival. Summary statistics, such as the total population size recorded for some years, contain little information about individual variation. Although in principle it is possible to separate the demographic and environmental components of the total annual change in population size, such an approach is not recommended because uncertainties in the estimates will be too large for practical applications. Consequently, efficient estimation of the demographic variance requires samples of individual fitnesses. Even a single year of fitness observations provides essential information about the demographic variance. The survival or death of a random sample of females must be recorded together with the number of female offspring each produces that survive to the next year. Let $w_1, w_2, \ldots, w_n$ denote such a sample of individual fitnesses in a particular year. The standard sum of squares statistic

$$\hat{\sigma}_d^2 = \frac{1}{n-1} \sum_{i=1}^{n} (w_i - \bar{w})^2 \qquad 1.10$$

gives an unbiased estimate of the demographic variance for the population size in the year the data are recorded (Sæther *et al.*, 1998a).

In general, the demographic variance may be a function of population size (Engen *et al.*, 1998), say $\sigma_d^2(N)$, and this relationship may be revealed by sampling individual fitness in different years with different population sizes, estimating the demographic variance in each year (see Figs 1.2 and 1.5). For example, if density dependence reduces both the mean and variance of individual fitness per year then the demographic variance will decrease with increasing population size. Alternatively, the demographic variance may be maximized at an intermediate population size if low (high) population density allows a high (low) proportion of adult females to survive and reproduce, because the variance of adult annual survival, $s(1-s)$, is a quadratic function of the mean survival, s.

The environmental variance can be estimated from a time series of population sizes in successive years, $N(1), N(2), \ldots, N(t)$. In a density-regulated population the expected multiplicative growth rate at a given time depends on the population size N as well as on parameters, collectively denoted by ψ, affecting the expected population growth. Letting $\Delta N(t) = N(t + 1) - N(t)$, the expected density-dependent multiplicative growth rate conditioned on the population size is given by $E[\Delta N/N | N] = \bar{\lambda}(N, \psi) - 1$. Assuming no temporal environmental autocorrelation and that $\Delta N/N$ depends on population size only in the previous year, the estimate of ψ, denoted as $\hat{\psi}$, is obtained by the method of least squares, minimizing the sum of the squared deviations between the observed and predicted sizes in the population time series (Stuart and Ord, 1999). This gives the corresponding estimate $\bar{\lambda}(N, \hat{\psi})$ of the expected density-dependent multiplicative growth rate.

The total stochastic variance causing fluctuations in population size, separated from changes due to density dependence, can be estimated from the residual variance of changes in population size not explained by population density. From equation 1.3 it follows that

$$E\left[\frac{\Delta N}{N} - \bar{\lambda}(N, \psi) + 1\right]^2 = \sigma_e^2 + \frac{\sigma_d^2}{N}.$$

This implies that the environmental variance at population size N in a given year can be estimated as

$$\hat{\sigma}_e^2(N) = \left[\frac{\Delta N}{N} - \bar{\lambda}(N, \hat{\psi}) + 1\right]^2 - \frac{\hat{\sigma}_d^2(N)}{N}. \qquad 1.11$$

Thus, after estimating the demographic variance and the expected multiplicative growth rate, an estimate of the environmental variance is available each year from the change in population size. Figure 1.5 depicts annual estimates of demographic and environmental variance plotted against population size for the Great Tit in Wytham Wood, assuming a logistic form of density regulation with a fixed carrying capacity (see eq. 2.7). If the environmental variance is approximately constant, independent of population size, then the estimate of σ_e^2 is simply the mean of the annual estimates. Table 1.2 gives estimates of σ_e^2 obtained by this procedure for populations in which an estimate of the demographic variance is available from data on individual survival and reproduction.

1.7.2 Environmental autocorrelation

The above method can be extended to allow for temporal environmental autocorrelation. For simplicity, we assume that the population size remains sufficiently large to neglect demographic stochasticity. The stochastic dynamics of population size can be written as

$$\frac{\Delta N(t)}{N(t)} = \bar{\lambda}(N(t), \psi) - 1 + \varepsilon(t)$$

where $\varepsilon(t)$ represents autocorrelated environmental noise with mean 0 and variance σ_e^2. The relatively short length of most population time series and the requirement for a small number of parameters to ensure statistical precision suggests the use of a

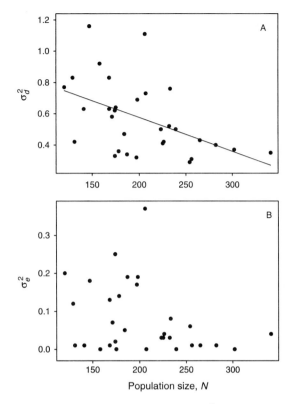

Fig. 1.5 Annual estimates of the demographic variance σ_d^2 and environmental variance σ_e^2 in multiplicative growth rate as a function of population size N for the Great Tit in Wytham Wood at Oxford, England.

simple form for the environmental autocorrelation function. We assume a geometric form for the environmental autocorrelation function, $\rho_e(\tau) = \kappa^{|\tau|}$, specified by the single parameter, κ.

Parameter estimation is facilitated by weighted differencing of the foregoing equation to convert the dynamics to a form with no autocorrelation in the noise,

$$\frac{\Delta N(t+1)}{N(t+1)} = \kappa \frac{\Delta N(t)}{N(t)} + \bar{\lambda}(N(t+1), \psi) - 1 - \kappa \bar{\lambda}(N(t), \psi) + \kappa + \xi(t)$$

where $\xi(t) = \varepsilon(t+1) - \kappa\varepsilon(t)$ has mean 0, variance $(1 - \kappa^2)\sigma_e^2$ and no autocorrelation. For a series of population sizes observed over L years the population parameters ψ, including the environmental variance σ_e^2 and the environmental autocorrelation parameter κ, determine the distribution of $(\Delta N(t+1)/N(t+1))$ conditional on population sizes in the two previous years, $f[(\Delta N(t+1)/N(t+1))|N(t+1), N(t), \psi]$, which is assumed to be normal. Parameter estimates are obtained by numerically maximizing the log-likelihood function

$$\ln \mathcal{L}(\psi) = \sum_{t=1}^{L-1} \ln f \left[\frac{\Delta N(t+1)}{N(t+1)} | N(t+1), N(t), \psi \right].$$

Estimates of the environmental autocorrelation parameter κ for several population time series are presented in Table 1.4.

The estimated environmental autocorrelation parameter κ is significantly negative for the Blue Tit and Mute Swan. There are several possible explanations for this. First, measurement errors in population estimates can cause successive values of the estimated multiplicative growth rate, $\lambda(t) = N(t+1)/N(t)$, to be negatively correlated. This occurs because the population size that appears in the denominator of $\lambda(t)$ also appears in the numerator of $\lambda(t-1) = N(t)/N(t-1)$, so that, for example, underestimation of $N(t)$ inflates $\lambda(t)$ and deflates $\lambda(t-1)$. The series for the Mute Swan probably has the lowest census accuracy of any of the data sets and is most likely to be influenced by this cause. Second, energy or resource limitation may cause individual females that reproduce many offspring one year to reproduce relatively few offspring the following year and *vice versa* (e.g. Tanaka *et al.*, 1970; Gustafsson and Sutherland, 1988). In large-bodied species with a long gestation and parental care, the average interbirth interval may be longer than 1 year (Owen-Smith, 1988). Such individual physiological effects could become synchronized at the population level by external environmental fluctuations. Third, negative maternal effects in reproduction between generations can create a similar outcome to such physiological effects within generations. For example, in mammals with multiple offspring per litter, such as mice, large-bodied mothers tend to produce many small offspring, and small mothers tend to produce few large offspring (Falconer, 1965; Kirkpatrick and Lande, 1989). In species with short generations, such as the Blue Tit that matures in a single year and has high adult mortality, such negative maternal effects could become synchronized in a population because of external environmental influences on body size. This explanation cannot apply to the species with long generations in Table 1.4, such as the Mute Swan. Fourth, time lags in population dynamics due to age structure create autocorrelation in population growth rates that could appear as environmental autocorrelation. However, age structure in species such as those in Table 1.4 typically produces population autocorrelations that are positive, not negative, as shown by the empirical population correlograms in Chapter 3 (Fig. 3.5). Finally, the external environment may actually be negatively autocorrelated, perhaps because of strong interaction with other species, such as food items, that have negative population autocorrelation.

Aside from the Mute Swan data set, which is most subject to sampling error, overall there is not much evidence of environmental autocorrelation, although the statistical power to detect this is weak in time series as short as most of those in Table 1.4. This provides some justification for the assumption of little or no environmental autocorrelation in subsequent models.

Table 1.4 Estimated environmental variance, $\hat{\sigma}_e^2$, with environmental autocorrelation function assumed geometric with parameter κ. Population size is assumed sufficiently large to neglect demographic stochasticity, as indicated by the mean population size, $\bar{N}$, and CV. Mean age at first breeding of females is α. Standard errors were obtained from 1,000 computer simulations of each time series using estimated population parameters. Significance of κ was tested by simulations using the population parameters estimated with $\kappa = 0$

Species	Locality	Study period	α	$\bar{N}$	CV	$\hat{\sigma}_e^2 \pm$ SE	$\hat{\kappa} \pm$ SE	Reference
Blue Tit	Wytham Wood, U.K.	1959–1995	1	297	0.28	0.036 ± 0.009	-0.37 ± 0.18*	R. McCleery (pers. com.)
Great Tit	Wytham Wood, U.K.	1960–1995	1	214	0.29	0.094 ± 0.023	-0.29 ± 0.25	R. McCleery (pers. com.)
Pied Flycatcher	Braunschweig, Germany	1964–1993	1	293	0.23	0.041 ± 0.011	-0.04 ± 0.26	Winkel (1996)
Grey Heron	Southern England	1928–1998	2	6719	0.18	0.011 ± 0.019	0.04 ± 0.18	Br. Tr. Ornith. (pers. com.)
Ibex, Rupicapra rupicapra	Swiss National Park	1944–1989	3	200	0.21	0.014 ± 0.007	-0.11 ± 0.20	F. Filli (pers. com.)
Moose, Alces alces	Isle Royale	1959–2001	3	1079	0.36	0.026 ± 0.008	0.07 ± 0.20	R. O. Peterson (pers. com.)
Mute Swan	River Thames, U.K.	1852–1939	4	500	0.21	0.015 ± 0.003	-0.38 ± 0.12**	Cramp (1972)

* Significantly different than 0 with probability $P < 0.05$.
** Significantly different than 0 with probability $P < 0.01$.

1.8 Summary

Population fluctuations in most species are produced by demographic and environmental stochasticity, rather than by internally driven cycles or chaos. This is because in most species the maximum expected rate of population growth is small, due to life history limitations and/or high density-independent mortality. Demographic stochasticity results from chance independent events of individual mortality and reproduction, causing random fluctuations in population growth rate, primarily in small populations. Environmental stochasticity results from temporal fluctuations in mortality and reproductive rates of all individuals in a population in the same or similar fashion, causing population growth rate to fluctuate randomly in populations of all sizes. In populations much larger than the ratio of the demographic variance to the environmental variance, demographic stochasticity can be neglected. The demographic variance can be estimated from data on individual survival and reproduction, and using this, the environmental variance can then be estimated from population time series.

Under density-independent growth in a random environment, the eventual rate of increase or decrease of population size is given by the long-run growth rate, the mean rate of increase of log population size, which is reduced by stochasticity. Density-dependent population growth in a stochastic environment produces temporal autocorrelation in population size, even in the absence of temporal autocorrelation in the environment. Time series for terrestrial populations of birds and mammals show little evidence of temporal environmental autocorrelation.

2

Extinction dynamics

2.1 Paleo-extinctions

Extinction is the eventual fate of all species. The vast majority of species that ever existed on earth during the past 2 billion years are now extinct. The mean longevity of species preserved in the fossil record is a few to several million years for different groups of animals (Table 2.1). This may overestimate the duration of all species because, in addition to having hard body parts, characteristics favoring preservation and discovery in the fossil record include high numerical abundance and large geographic range, both of which are associated with increased species longevity during background extinction (Jablonski, 1989, 1991, 1994).

Major and minor mass extinctions in the fossil record are used to define the boundaries of geological periods. There have been five major mass extinctions in earth's history that together account for about 5–10% of all paleontological extinctions (Jablonski, 1986; Erwin, 2001). These are illustrated for changes in the diversity of marine invertebrate families during the past 600 million years in Figure 2.1 and Table 2.2. The most recent occurred 65 million years ago at the end of the Cretaceous period due to an asteroid impact off the Yucatan peninsula (Sharpton *et al.*, 1992, 1993) with a physical force and ecological effects comparable with what might occur in a nuclear world war. In a mass extinction the percentage of extinctions increases at lower taxonomic levels because higher taxa that survive a mass extinction often lose a substantial fraction of their subtaxa in the event. For example, the end Cretaceous event extinguished 16% of families, 47% of genera, and an estimated 76% of species of marine invertebrates then existing on earth (Table 2.2). It also extirpated the dinosaurs, and opened the way for the adaptive radiation of mammals (Romer, 1966; Carroll, 1988). Other major and minor mass extinctions may have been caused by asteroid impacts or natural climatic and ecological changes.

Table 2.1 Mean durations of species in the fossil record in millions of years (Myr)

Taxon	Mean duration (Myr)	Reference
Eocene–Pleistocene mammals	2.6	Alroy (2000)
Echinoderms	6	Durham (1970)
Cenozoic bivalves	10	Raup and Stanley (1978)
Cenozoic planktonic foraminifera	8–9	Parker and Arnold (1997)

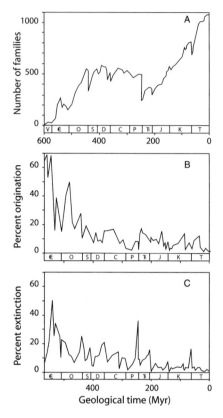

Fig. 2.1 Diversity of marine invertebrate families during the past 600 million years. (From Sepkoski, 1993, 1995.)

Table 2.2 Percent extinctions of marine invertebrates at different taxonomic levels in the five major mass extinctions. (From Sepkoski, 1993, 1995; Jablonski, 1994.)

Mass extinction (age)	Families[1]	Genera[1]	Species[2]
End Ordovician (439 Myr)	26	60	85
Late Devonian (367 Myr)	22	57	83
End Permian (245 Myr)	51	82	95
End Triassic (208 Myr)	22	53	80
End Cretaceous (65 Myr)	16	47	76

[1] Observed in fossil record.
[2] Estimated by reverse rarefaction of species within genera.

The largest mass extinction occurred 245 million years ago at the end of the Permian period, terminating an estimated 95% of marine invertebrate species that existed then (Jablonski, 1991, 1994). Species involved in mass extinctions appear to be largely random, although some major taxonomic groups are more affected than others.

The majority of all extinctions in the fossil record occurred as background extinctions at more normal rates (Jablonski, 1986, 1994). For taxa that perished in background extinctions the distribution of taxon longevity is roughly exponential at all taxonomic levels, including families, genera, and species (Van Valen, 1973, 1979; Wei and Kennet, 1983; Parker and Arnold, 1997). A roughly exponential distribution of taxon longevity allows some fluctuations in extinction rate per unit time, but implies a stochastically constant risk of extinction per increment of taxon age. Exponential distributions of taxon longevity have been interpreted by many paleontologists as indicating that background extinctions are largely random. However, the apparent randomness is largely due to our ignorance about the detailed causes of background extinctions. Because higher taxa (above the species level) differ substantially in the number of subtaxa they include (e.g. Sepkoski, 1978, fig. 11), exponential distributions of longevity for higher taxa imply that related subtaxa must be subject to similar extinction risks. For example, if a large, species-rich genus has the same risk of extinction per increment of age as a monotypic genus, then species in the large genus must share a greatly elevated risk of extinction compared with the single species in the monotypic genus.

2.2 Modern anthropogenic extinctions

Extinction is an important natural process that can facilitate the evolution of new species and higher taxa. Tragically, human activities have greatly accelerated the rate of species extinctions and are now causing the sixth major mass extinction in the earth's history. Judging from the fossil record after previous major mass extinctions (e.g. Fig. 2.1), it will take about 5–10 million years for biodiversity to recover, even without continued human interference (Jablonski, 1991, 1994; Erwin, 2001). The current mass extinction is thought to have begun about 11,000 years ago with increasing hunting by an expanding human population contributing to the extinction of dozens of genera of large mammals at the end of the Pleistocene period in North America and Europe, including the Mastodon *Mammut* spp. and Wooly Mammoth *Mammuthus* spp. (Martin and Klein, 1984; Martin, 1986; MacPhee, 1999; Alroy, 2001). Figure 2.2 plots origination and extinction rates of large-bodied and small-bodied mammalian species during the late Pleistocene, showing natural background rates of extinction and the end-Pleistocene anthropogenic catastrophe affecting large mammals. Similarly, human colonization of the Hawaiian islands and other Pacific islands a few thousand years ago led to the extinction of a high proportion of the relatively large and/or flightless bird species (James, 1995; Steadman, 1995). Because of their limited population sizes and geographic ranges, endemic island species are

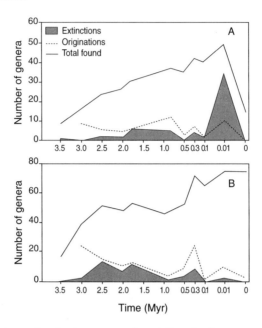

Fig. 2.2 Origination and extinction rates, and total diversity for genera of North American mammals during the late Pleistocene, shown separately for species with (A) large (>44 kg) and (B) small (<44 kg) adult body weights. Note change of time scale in the last 0.5 Myr. (From Martin and Steadman, 1999.)

especially vulnerable to anthropogenic extinction from hunting, introduced species, and habitat alteration (Martin, 1984; Groombridge, 1992; Case *et al.*, 1998).

Anthropogenic extinctions accelerated during recorded human history in parallel with an increasing human population (Groombridge, 1992; Cohen, 1995). About one-eighth of the avian species and one-quarter of the mammalian species of the world are now threatened with extinction due to human activities, primarily habitat destruction and fragmentation, overexploitation, introduced species, and pollution (Groombridge, 1992; Hilton-Taylor, 2000). The number of species recently living on earth is on the order of 10 million or more, with only about 2 million having been scientifically described (Groombridge, 1992; Heywood, 1995). It has been estimated using biogeographic theory that anthropogenic extinctions are currently causing the loss of a few to several percent of this biodiversity per decade, mostly in species-rich tropical forests and coral reefs (Groombridge, 1992; May *et al.*, 1995; Bryant *et al.*, 1998). This is about three or four orders of magnitude greater than the rate of natural background extinctions (May *et al.*, 1995; Pimm *et al.*, 1995; McKinney, 1998). Further details on threatened and endangered species appear in Chapters 5 and 6.

Most modern extinctions are largely attributable to human causes that accelerated the demise of these species. Population trajectories for some well-documented

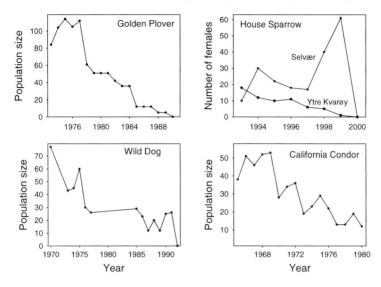

Fig. 2.3 Decline to extinction of a Scottish population of Golden Plover *Pluvialis apricaria* (Parr, 1992), the Wild Dog *Lycaon pictus* in Serengeti National Park, Tanzania (Burrows *et al.*, 1995), two populations of House Sparrow *Passer domesticus* at Helgeland, northern Norway (Sæther *et al.*, unpubl.), and the California Condor *Gymnogyps californianus* (Dennis *et al.*, 1991).

extinctions are presented in Figure 2.3. These show that recent extinctions of vertebrate populations and species usually occur through irregular population declines lasting many years, with the final extinction happening from a small population size. However, some small populations are extinguished by natural or artificial catastrophes, such as the island population of House Sparrows illustrated in Figure 2.3, which became suddenly extinct in a single year, probably due to disease and/or immigration of a pair of Sparrow Hawks (*Accipter nisus*) to the island. Stochasticity can cause or accelerate extinctions, but in many cases anthropogenic influences make it impossible for populations to sustain themselves, and stochasticity is merely superimposed on deterministic declines (Caughley, 1994), such as for the California Condor in Figure 2.3.

2.3 Diffusion approximation of stochastic dynamics

This chapter concerns the dynamics of population extinction, especially the behavior of trajectories in small populations, the mean time to extinction, and the expected duration of the final decline to extinction. Several approaches have been taken to analyze such problems. Methods with a discrete state space that preserve integer values of population dynamics include continuous-time branching processes and birth-and-death processes, and discrete-time transition matrix iteration. Branching processes and birth-and-death processes are appropriate to analyze density-dependent population growth with demographic stochasticity, but generally do not include environmental stochasticity (MacArthur and Wilson, 1967; Goel and Richter-Dyn, 1974; Karlin and Taylor, 1975; Nisbet and Gurney, 1982; Renshaw, 1991; Nåsell, 2001). Markov model transition matrices describe density-dependent population growth in discrete time by specifying the probabilities per unit time of transitions between all possible pairs of population sizes, and can include both demographic and environmental stochasticity, as well as large catastrophes (Mangel and Tier, 1993). However, transition matrix models generally involve extensive numerical calculation by repeated matrix multiplications, which for populations of moderate or large size becomes time consuming and potentially inaccurate due to accumulation of rounding errors. The heavy reliance on numerical computation in transition matrix models limits their ability to provide general insights, and they appear to have few advantages over the relatively straightforward simulation of stochastic population dynamics.

One of the most general and powerful methods for analyzing stochastic population dynamics is provided by diffusion approximations (Cox and Miller, 1965; Karlin and Taylor, 1981). A stochastic population model in discrete time can be approximated by a diffusion process that is continuous in both time and population size, driven by white noise (with no temporal autocorrelation). Sample paths of diffusion processes are likely to change by a small proportion of the usual range of population sizes in a short time interval, although they may occasionally undergo rapid substantial excursions. Diffusion theory for one-dimensional processes is well suited to describe the stochastic dynamics of single populations with no age structure, little or no environmental autocorrelation, and that usually change by small increments. In Chapter 1 we cited evidence against the frequent occurrence of intrinsic cycles and chaos, including the gradual declines to extinction depicted in Figure 2.3. Catastrophic events that suddenly reduce population size by a substantial fraction are occasionally observed in modern populations as a result of disease epidemics and extreme climatic effects such as drought (Young, 1994). We will show below that the mean time to extinction under random catastrophes scales with population carrying capacity like that for continual environmental stochasticity, provided that the population can not be extinguished by a single catastrophe. This supports the view that catastrophes can be considered as extremes of environmental stochasticity (Shaffer, 1987; Caughley, 1994; Erb and Boyce, 1999) when calculating the mean time to extinction. We will also describe how short-term environmental autocorrelation can be accommodated in a diffusion approximation.

The theory of diffusion processes involves some subtle and difficult concepts (Karlin and Taylor, 1981). Fortunately, it is often straightforward to derive a diffusion approximation to a stochastic model and to obtain important results using relatively simple methods. We first explain how to derive a diffusion approximation to a stochastic model in discrete or continuous time, and then apply this to the analysis of population fluctuations and extinction dynamics.

2.3.1 Infinitesimal mean and variance

A general stochastic difference equation can be expressed in terms of the change of population size per unit time, $\Delta N(t) = N(t + 1) - N(t)$, using the notation $N \equiv N(t)$,

$$\Delta N = M(N) + \sqrt{V(N)}\varepsilon(t).$$ 2.1a

The discrete-time random variable $\varepsilon(t)$ is defined to have zero mean, $\bar{\varepsilon} = 0$, unit variance, $\sigma_\varepsilon^2 = 1$, and no temporal autocorrelation. Here $M(N)$ represents the expected change and $V(N)$ represents the variance of change in population size per unit time, for a given population size at time t,

$$M(N) = \mathrm{E}[\Delta N|N] \quad \text{and} \quad V(N) = \mathrm{Var}[\Delta N|N].$$ 2.1b

A diffusion process is characterized by the mean and variance of the change in a random variable in an infinitesimal time interval, as well as initial and boundary conditions. The simplest way of deriving the infinitesimal mean and variance of a diffusion approximation for a population reproducing at discrete-time intervals is to obtain them directly from the mean and variance of the change per unit time in the stochastic difference equation 2.1a. The infinitesimal mean, $M(N)$, also called the drift coefficient, gives the expected behavior of population trajectories over a short time and represents the predictable or deterministic component of the process. The infinitesimal variance, $V(N)$, also called the diffusion coefficient, gives the variance of population trajectories over a short time and represents the stochastic or unpredictable component, assuming no temporal autocorrelation of stochastic factors. These appear in the diffusion equation, also known as the Kolmogorov forward (or Fokker–Planck) equation, governing the dynamics of the probability density function for population size at time t, $\phi(N, t)$,

$$\frac{\partial \phi}{\partial t} = -\frac{\partial}{\partial N}[M(N)\phi] + \frac{1}{2}\frac{\partial^2}{\partial N^2}[V(N)\phi].$$ 2.1c

This partial differential equation is difficult to solve except in elementary cases. Nevertheless, a great deal of useful information about a diffusion process can be derived from relatively simple analysis of its infinitesimal mean and variance, as shown below.

As our first example, consider the basic discrete-time model of density-independent growth in a random environment with no environmental autocorrelation, $\Delta N = [\lambda(t) - 1]N$. The infinitesimal mean and variance of the diffusion approximation can be written, using equation 1.3, as

$$M(N) = (\bar{\lambda} - 1)N \quad \text{and} \quad V(N) = \sigma_e^2 N^2 + \sigma_d^2 N.$$ 2.2a

2.3.2 Scale transformation

For purposes of mathematical or statistical analysis it is often convenient to describe population dynamics on a transformed scale of measurement. A diffusion process on a new scale of measurement for population size, defined by the monotonic transformation $X = g(N)$, can by derived from the transformation formulas for the infinitesimal mean and variance obtained by Taylor series expansion of $g(N)$ (Karlin and Taylor, 1981),

$$\mu(X) = g'(N)M(N) + \tfrac{1}{2}g''(N)V(N) \quad \text{and} \quad \nu(X) = [g'(N)]^2 V(N)$$

where a prime $'$ indicates a derivative with respect to N. In these formulas N must finally be expressed as a function of X using the inverse transformation $N = g^{-1}(X)$. Note that on the transformed scale the infinitesimal mean representing the deterministic component is modified by stochasticity in the transformation. This implies that the separation of deterministic and stochastic components depends on the scale of measurement for population size.

For example, transforming the infinitesimal moments for density-independent growth in equation 2.2a to the natural log scale, $X = \ln N$, we have $g(N) = \ln N$, $g'(N) = 1/N$, and $g''(N) = -1/N^2$. This produces $\mu(X) = \bar{\lambda} - 1 - \nu(X)/2$ and $\nu(X) = \sigma_e^2 + e^{-X}\sigma_d^2$. Comparison with the discrete-time formulas (eqs. 1.3, 1.4) indicates that the diffusion approximation is accurate when $\lambda(t)$ has a small or moderate coefficient of variation and $\bar{\lambda}$ is close to one, $|\bar{\lambda} - 1| \ll 1$. For population sizes large enough to neglect demographic stochasticity, the infinitesimal mean gives the diffusion approximation for the long-run growth rate, $\bar{r} = \bar{\lambda} - 1 - \sigma_e^2/2$.

If instead we first transform the discrete-time stochastic dynamics $N(t + 1) = \lambda(t)N(t)$ to the log scale by taking logs of both sides, giving $\Delta X = r(t)$, and then derive the infinitesimal mean and variance, we obtain a different diffusion approximation. Neglecting demographic stochasticity, this produces the infinitesimal mean and variance as constants, $M(X) = \bar{r}$ and $V(X) = \sigma_r^2$. A diffusion process with constant infinitesimal mean and variance represents Brownian motion with drift, also known as a Wiener process (Cox and Miller, 1965; Karlin and Taylor, 1981) producing a Gaussian distribution for $X(t)$, which in this case has mean and variance exactly as in the discrete time model (eqs 1.6b).

As another example, under stochastic growth with deterministic density regulation of the Gompertz form, the discrete-time model on the log scale is $\Delta X = r(t) - \gamma X$. Neglecting demographic stochasticity, the diffusion approximation has infinitesimal mean and variance

$$M(X) = \bar{r} - \gamma X \quad \text{and} \quad V(X) = \sigma_r^2. \qquad 2.2b$$

A diffusion process with a linear infinitesimal mean and constant infinitesimal variance represents an Ornstein–Uhlenbeck process (Cox and Miller, 1965; Karlin and Taylor, 1981) producing a Gaussian distribution of $X(t)$. Differential equations and solutions for the mean and variance of $X(t)$ can be obtained from those for the discrete-time process (eqs 1.9a) by letting the time step become small, $\Delta t \to dt$, yielding

$$\bar{X}(t) = \frac{\bar{r}}{\gamma} + \left[X_0 - \frac{\bar{r}}{\gamma}\right]e^{-\gamma t} \quad \text{and} \quad \sigma_{\bar{X}}^2(t) = \frac{\sigma_r^2}{2\gamma}(1 - e^{-2\gamma t}) \qquad 2.2c$$

and population autocorrelation function $\rho(\tau) = e^{-\gamma|\tau|}$. Comparison with the corresponding discrete-time formulas (eqs 1.9b) indicates that the diffusion approximation is most accurate under weak density regulation, $\gamma \ll 1$. These examples confirm the accuracy of diffusion approximations for populations that are expected to change by only a small proportion of the usual range each year.

2.3.3 Stochastic differential equations

In complex models, such as for spatially distributed populations (Chapter 4) or multiple interacting species (Chapter 8), the analysis may be facilitated by introducing stochasticity directly into a continuous-time model. The analog of the stochastic difference equation 2.1a in continuous-time is the stochastic differential equation

$$\frac{dN}{dt} = M(N) + \sqrt{V(N)}\frac{dB(t)}{dt} \quad \text{or} \quad dN = M(N)dt + \sqrt{V(N)}dB(t)$$

The continuous-time analog of the discrete-time random variable $\varepsilon(t)$ in equation 2.1a is 'white noise' with zero mean and no temporal autocorrelation, represented by the increment of standard Brownian motion $dB(t)$ occurring in an infinitesimal time increment dt. Starting from a given initial position at time $t = 0$ standard Brownian motion $B(t)$ describes a continuous-time random walk generating a probability density function that is normal with expected value at the initial position and variance t. In the infinitesimal time increment dt the increment of the variance for standard Brownian motion therefore equals dt, which implies that $E[dB(t)] = 0$ and $Var[dB(t)] = dt$ (Karlin and Taylor, 1981, p. 347).

To obtain the infinitesimal mean and variance of a diffusion process corresponding to a stochastic differential equation, itself taken as an approximation to an underlying discrete-time model, following Turelli (1977) we use the Ito stochastic calculus, which is consistent with the standard rules of statistics (Karlin and Taylor, 1981). The infinitesimal mean and variance of the diffusion process described by the stochastic differential equation can then be defined as

$$M(N) = \frac{1}{dt}E[dN|N] \quad \text{and} \quad V(N) = \frac{1}{dt}Var[dN|N].$$

2.3.4 Temporally autocorrelated environment

A diffusion approximation can incorporate temporal autocorrelation in the environment, provided that the time scale for the event of interest, such as extinction, is much longer than the time scale of environmental autocorrelation. Comparison of the solutions for the discrete-time models of density-independent population growth on the log scale in uncorrelated and autocorrelated environments (eqs 1.6, 1.8) suggests that, in the diffusion approximation to the model with environmental autocorrelation, in the infinitesimal variance the environmental variance should be multiplied by the total environmental autocorrelation over all time lags from $-\infty$ to ∞, that is $1 + 2\sum_{\tau=1}^{\infty} \rho_e(\tau)$, without changing the infinitesimal mean (Turelli, 1977; Lande, 1993; Foley, 1994, 1997; Lande et al., 1995). Evidently, positive temporal autocorrelation magnifies environmental stochasticity, whereas negative temporal autocorrelation diminishes it. This also applies in the presence of demographic stochasticity and density dependence. If the original model is a continuous-time stochastic differential equation, then in the infinitesimal variance the environmental variance should be multiplied by the integral of the environmental autocorrelation function, $\int_{-\infty}^{\infty} \rho_e(\tau)\,d\tau$. The scale transformation formulas (after eq. 2.2a) assume no environmental autocorrelation and should not be applied in its presence.

2.4 Extinction trajectories in small populations

In models of small populations well below carrying capacity, density dependence may be neglected, but demographic stochasticity should be included along with environmental stochasticity, because the interaction of these two stochastic factors can strongly influence the dynamics of small populations. We have seen in Chapter 1 that environmental stochasticity reduces the long-run growth rate of a population at all sizes. Because demographic stochasticity operates more strongly on smaller populations, it exerts an increasingly strong downward trend on trajectories at smaller population sizes (eqs 1.3 and 1.4).

The behavior of sample paths in density-independent populations subject to both demographic and environmental stochasticity can be analyzed by applying a scale transformation that renders the infinitesimal variance constant, independent of N (Lande, 1998). A constant infinitesimal variance on the transformed scale guarantees that the sign of the infinitesimal mean directly reveals the tendency of sample paths to increase or decrease, as does the long-run growth rate in the basic model of density-independent growth in a random environment (see Chapter 1.5.1). Using the transformation formulas for the infinitesimal mean and variance (Chapter 2.3.2) the scale transformation $X = \int^{N} \left[1/\sqrt{V(z)}\right] dz$ produces a diffusion process with a new infinitesimal mean and a constant infinitesimal variance,

$$\mu(X) = \frac{M(N) - V'(N)/4}{\sqrt{V(N)}} \quad \text{and} \quad \nu(X) = 1$$

where again a prime ′ indicates a derivative. On this scale a stochastically unstable equilibrium can occur corresponding to a population size N^* on the original scale that is the solution of $\mu(X) = 0$ or $M(N^*) = V'(N^*)/4$. Using the infinitesimal moments for density-independent growth in equations 2.2a yields

$$N^* = \frac{\sigma_d^2/4}{\bar{\lambda} - 1 - \sigma_e^2/2}. \qquad 2.3$$

For populations with a positive long-run growth rate an unstable equilibrium occurs on the new scale corresponding to a population size of a quarter the demographic variance divided by the long-run growth rate (see Chapter 2.3.2). Populations below this size tend to decline rapidly to extinction as a result of demographic stochasticity. Figure 2.4 portrays simulated trajectories in small populations undergoing density-independent growth subject to the joint operation of demographic and environmental stochasticity. Figure 2.5 depicts the infinitesimal mean on the transformed scale for the identical model, showing how demographic stochasticity creates an unstable equilibrium on the transformed scale. These results illustrate the importance of demographic stochasticity in contributing to extinction risk, not only by increasing the fluctuations

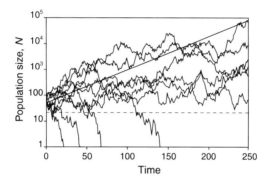

Fig. 2.4 Simulated trajectories for stochastic density-independent population growth, $N(t+1) = \lambda(t)N(t)$, with demographic variance $\sigma_d^2 = 1$, environmental variance $\sigma_e^2 = 0.04$, and expected multiplicative growth rate $\bar{\lambda} = 1.03$. Solid line gives the deterministic dynamics of geometric growth at the expected rate $\bar{\lambda}$. Horizontal dashed line at $N^* = 25$ corresponds to the unstable equilibrium on the transformed scale (eq. 2.3). Populations that fall below N^* tend to become rapidly extinct because of demographic stochasticity. (From Lande, 2002.)

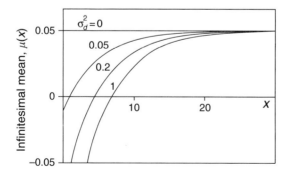

Fig. 2.5 Infinitesimal mean of a diffusion process, $\mu(X)$, on the transformed scale, X, with unit infinitesimal variance, for different values of the demographic variance, σ_d^2. The dynamics of population size on the original scale, N, are otherwise the same as in Figure 2.4. (For further details see Lande, 1998.)

in small populations but also by producing a strong tendency for small populations to decline.

Populations reduced to small size often experience deterministic declines in mean individual fitness due to the failure of cooperative interactions among individuals, such as group foraging, group defense against predators, communal nesting, social facilitation of reproduction, or most generally the difficulty of finding a mate in a small or sparsely distributed population. This is known as an Allee effect or depensation (Allee *et al.*, 1949; Dennis, 1989, 2002; Courchamp *et al.*, 1999; Stephens *et al.*, 1999). For example, sparse populations of sea urchins suffer reduced fertilization success during broadcast spawning of gametes from separate sexes (Levitan and Petersen, 1995). Similarly, small populations of self-incompatible monoecious plants are less attractive to pollinators, which reduces their seed set (Groom, 1998). Strong Allee effects can render the expected multiplicative growth rate of a small population less than 1, creating an unstable equilibrium at small population size, below which a population tends to decline deterministically to extinction.

Population genetic processes may produce results resembling an Allee effect. In most animal species the sex of an individual is determined by random Mendelian segregation of X and Y chromosomes (Bull, 1983); this produces random fluctuations in sex ratio that reduce the long-run growth rate of small populations, especially with a strictly monogamous breeding system (Engen *et al.*, 2003a). Random genetic drift in small populations tends to cause a loss of genetic variation, diminishing the potential for evolutionary adaptation to a changing environment (Wright, 1931; Franklin, 1980; Lande, 1995). Matings between closely related individuals in small populations lead to homozygosity of recessive deleterious mutations and inbreeding depression in fitness (Falconer and MacKay, 1996; Lynch and Walsh, 1998). New mildly deleterious mutations can become fixed by random genetic drift in a small population, gradually eroding its fitness (Lande, 1995).

As shown by Figures 2.4 and 2.5, the operation of demographic stochasticity closely resembles an Allee effect, and distinguishing their influence using data on the trajectory of a small population may be difficult (Lande, 1998). Distinguishing demographic stochasticity from a classical Allee effect would require detailed investigations of ecological, behavioral, and genetic mechanisms underlying the dynamics of small populations (Courchamp *et al.*, 1999; Stephens *et al.*, 1999).

2.5 Stationary distribution of population size

Neglecting demographic stochasticity, in the diffusion approximation to a population model $N = 0$ generally is a natural (inaccessible) boundary (Karlin and Taylor, 1981) and with density limitation of the population a stationary distribution of population size may exist. This can be obtained by solving the diffusion equation 2.1b for the equilibrium distribution approached for long times, $\phi(N) \equiv \phi(N, \infty)$, so that $\partial \phi / \partial t = 0$, which produces an ordinary differential equation, $\frac{1}{2}[V(N)\phi(N)]'' - [M(N)\phi(N)]' = 0$. This can be integrated by elementary methods: a first integration and division by $V(N)\phi(N)/2$, followed by a second integration assuming that $\phi(N) \to 0$ as $N \to \infty$, requires that the first constant of integration be 0, giving finally

$$\phi(N) = \frac{const}{V(N)} \exp\left\{2 \int^N \frac{M(x)}{V(x)} dx\right\} \qquad 2.4$$

where *const* is a normalization constant such that $\int_0^\infty \phi(N)\, dN = 1$ (Wright, 1931; Karlin and Taylor, 1981). A stationary distribution does not exist when it cannot be normalized; most commonly this occurs when the functions in equation 2.4 blow up for $N \to 0$. For example, considering small population sizes well below carrying capacity, we can employ the density-independent model with demographic stochasticity (eq. 2.2a) to show that for $N \ll \sigma_d^2/\sigma_e^2$ formula 2.4 is approximately proportional to $1/N$, which cannot be integrated over any range of population sizes that includes 0.

2.6 Quasi-stationary distribution and mean time to extinction

Extinction is the eventual fate of all populations, as suggested by the fossil record. In the present models extinction inevitably occurs because with density limitation $N = \infty$ is a natural (inaccessible) boundary, and with demographic stochasticity $N = 0$ is an absorbing boundary (Karlin and Taylor, 1981). Thus with demographic stochasticity a true stationary distribution does not exist. Nevertheless, for populations with a positive long-run growth rate and large carrying capacity, extinction may be rather unlikely to occur for a long time and the population may fluctuate in a quasi-stationary distribution that decays slowly with time.

Questions about the time required for a stochastic process with a given initial value to exit a certain interval are called first passage time (or exit) problems. These can be addressed most accurately using the Kolmogorov backward equation (Karlin and Taylor, 1981) governing the probability density function $\Xi(N_0, t)$ of the exit time, t, given an initial size N_0,

$$\frac{\partial \Xi}{\partial t} = M(N_0)\frac{\partial \Xi}{\partial N_0} + \frac{V(N_0)}{2}\frac{\partial^2 \Xi}{\partial N_0^2}. \qquad 2.5a$$

We impose a lower absorbing boundary at population size $C \le N_0$, which may be set at extinction, $C = 0$, or some larger size of interest. Some authors define quasi-extinction as population decline to a small level at which there is a serious risk of extinction in the short term (see below, and Ginzburg *et al.*, 1982). Infinity generally constitutes a natural boundary that is inaccessible in population models with realistic density dependence, and hence ∞ can be treated as an absorbing barrier that actually is never reached.

The mean time to extinction, or more generally to reach the lower boundary C, can be defined as $T(N_0) = \int_0^\infty t\, \Xi(N_0, t)\, dt$. Multiplying both sides of equation 2.5a by t and integrating from $t = 0$ to ∞ produces an ordinary differential equation for the mean time to extinction,

$$M(N_0)\frac{\partial T}{\partial N_0} + \frac{V(N_0)}{2}\frac{\partial^2 T}{\partial N_0^2} = -1 \qquad \text{2.5b}$$

where the right side is obtained using integration by parts with $\Xi(N_0, \infty) = 0$ and $\int_0^\infty \Xi(N_0, t)\,dt = 1$. The absorbing boundary conditions are $T(C) = T(\infty) = 0$ because no time is required to reach a boundary when starting from one of them.

The expected cumulative time spent at each population size before first reaching C is the Green function, or sojourn time, for the diffusion process,

$$G(N, N_0) = \begin{cases} \dfrac{2S(N)}{s(N)V(N)} & \text{for } C \le N \le N_0 \\[2ex] \dfrac{2S(N_0)}{s(N)V(N)} & \text{for } N \ge N_0 \end{cases} \qquad \text{2.6a}$$

where

$$s(N) = \exp\left\{-2\int_C^N \frac{M(x)}{V(x)}\,dx\right\} \quad \text{and} \quad S(N) = \int_C^N s(x)\,dx$$

(Karlin and Taylor, 1981; Lande *et al.*, 1995). For populations with a positive long-run growth rate below carrying capacity, K, it can be shown that the Green function is nearly independent of N_0 for an initial population size in the vicinity of K. This is because such a population, starting anywhere above a small size, is likely to grow to carrying capacity and spend a long time fluctuating around it.

The mean time for the population size to first reach C starting from $N_0 \ge C$ is the integral of the Green function over all population sizes between the boundaries at C and ∞, and changing the order of integration produces

$$T(N_0) = \int_C^\infty G(N, N_0)\,dN = 2\int_C^{N_0} s(x)\int_x^\infty \frac{1}{s(N)V(N)}\,dN\,dx \qquad \text{2.6b}$$

as previously derived by Goel and Richter-Dyn (1974) and Leigh (1981) for $C = 0$.

The probability distribution of population size starting from N_0 before finally reaching extinction is $G(N, N_0)/T(N_0)$. This is called the quasi-stationary distribution because with a large carrying capacity, K, and a positive long-run growth rate below K, populations starting in the vicinity of K are expected to spend a long time fluctuating around K in a nearly stationary distribution before finally becoming extinct. Note that the quasi-stationary distribution of population size is proportional to the formula for the stationary distribution above the initial population size (compare eq. 2.4 with eq. 2.6a).

Examples of quasi-stationary and stationary distributions of population size are depicted in Figure 2.6 for a discrete-time logistic model characterized by linear density dependence, with stochastic density-independent multiplicative rate $\lambda(t) = 1 + \beta(t)$ and stochastic carrying capacity $K(t)$,

$$\Delta N = \beta(t)N - \frac{\bar{\beta}}{K(t)}N^2. \qquad 2.7$$

We incorporate demographic and environmental stochasticity in density-independent growth rate but only environmental stochasticity in the carrying capacity. Environmental noise with no temporal autocorrelation, $\varepsilon(t)$, is added to the logarithm of carrying capacity, $K(t) = \bar{K}e^{-\varepsilon(t)}$ such that $\sigma_{\ln K}^2 = \sigma_\varepsilon^2$. The joint distribution of $\beta(t)$ and $\varepsilon(t)$ is assumed to be bivariate normal with environmental correlation ρ. We assume $\bar{\varepsilon} = -\sigma_\varepsilon^2/2$ to guarantee that fluctuations in carrying capacity do not alter the expected dynamics (see after eq. 1.5b). Thus $e^{-\varepsilon(t)}$ and $e^{\varepsilon(t)}$ have identical distributions, with $E[K(t)] = \bar{K}$ and $E[1/K(t)] = 1/\bar{K}$, and for any given positive value of $\beta(t)$ the carrying capacity has a lognormal distribution with a squared coefficient of variation of $e^{\sigma_{\ln K}^2} - 1$ (Johnson and Kotz, 1970). This formulation of environmental stochasticity in carrying capacity is more realistic than previous models (reviewed by Feldman and Roughgarden, 1975), which not only allowed stochasticity to alter the

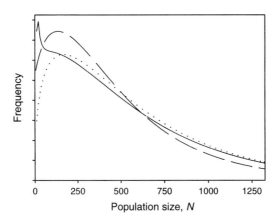

Fig. 2.6 Stationary distribution ($\sigma_d^2 = 0$, solid line) and quasi-stationary distribution ($\sigma_d^2 = 1$, dashed line) of population size in the logistic model (eq. 2.7) with stochastic density-independent growth rate and deterministic density dependence ($\sigma_{\ln K}^2 = 0$). Other parameters are $K = 1000, \bar{\beta} = 0.06, \sigma_\varepsilon^2 = 0.05$. Also shown is the quasi-stationary distribution of population size with the same parameters and in addition stochasticity in carrying capacity ($\sigma_{\ln K}^2 = 2$ and $\rho = 0$, dotted line).

expected dynamics but also permitted negative carrying capacities. The infinitesimal mean and variance of the diffusion approximation are

$$M(N) = \bar{\beta} N \left(1 - \frac{N}{\bar{K}} \right)$$

$$V(N) = \sigma_d^2 N + \sigma_e^2 N^2 + 2\rho \sigma_e \sigma_{\ln K} \frac{\bar{\beta}}{\bar{K}} N^3 + \left(e^{\sigma_{\ln K}^2} - 1 \right) \left(\frac{\bar{\beta}}{\bar{K}} \right)^2 N^4.$$

2.6.1 Estimating the quasi-stationary distribution

To calculate the quasi-stationary distribution of a real population it is necessary to assume a particular dynamic model to estimate the population parameters. For this purpose we employed data on individual fitness and population fluctuations in a small population of Song Sparrows (Sæther *et al.*, 2000a). The form of density dependence was found to be nearly logistic by fitting the more general theta-logistic model with deterministic density dependence (eq. 5.1). Data were fit to the model by the estimation method at the end of Chapter 1. Parameter estimates were used in equations 2.6a,b to calculate the quasi-stationary distribution of population size. Similar estimates were performed for populations of Pied Flycatcher and Soay Sheep (Fig. 2.7). Displacement of the population sizes observed over a limited timespan from the predicted quasi-stationary distribution can be attributed to (a) sampling error in estimates of population parameters used to calculate the quasi-stationary distribution, and (b) temporal autocorrelation in population dynamics that can produce a substantial sampling error in the observed mean population size (see after eq. 3.11).

2.6.2 Asymptotic scaling of mean time to extinction

Management of declining populations often is aimed at protecting or restoring habitat to increase K. Investigating how the mean time to extinction depends on the carrying capacity can help to decide over what range of carrying capacities different management actions might produce major increases or diminishing returns in population longevity.

A general qualitative view of the relative risks of population extinction posed by different stochastic factors can be obtained by deriving how the mean time to extinction starting from population carrying capacity, K, scales as a function of K under each factor alone. Lande (1993) investigated this question by including single stochastic factors in a simple deterministic continuous-time model of density-independent growth up to a ceiling or reflecting boundary at K. For consistency with the present exposition, we consider here the deterministic discrete-time density-independent model $\Delta N = (\lambda - 1)N(t)$ with a ceiling at K. Demographic and environmental stochasticity in λ can be analyzed using diffusion approximations. Catastrophes that suddenly reduce log population size by an amount ε at random times with expected frequency f can be analyzed using a delay-differential equation

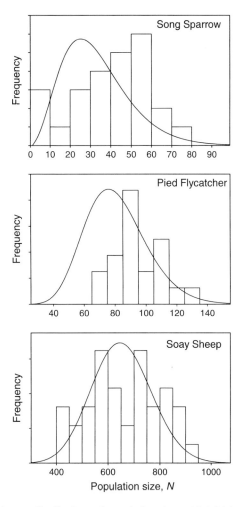

Fig. 2.7 Quasi-stationary distributions of population size, with initial population size at carrying capacity, $N_0 = K$, fit to the theta-logistic model with stochastic density-independent growth rate (eq. 5.1). Song Sparrow on Mandarte Island, B.C., Canada, with parameter estimates $\hat{\beta}_1 = 0.99$, $\hat{K} = 41$, $\hat{\theta} = 1.09$, $\hat{\sigma}_e^2 = 0.41$, $\hat{\sigma}_d^2 = 0.66$. Bars show the observed distribution of population size during a 24-year period (Fig. 5.2). Pied Flycatcher at Hoge Veluwe, central Netherlands (Sæther *et al.*, 2002a), with parameter estimates $\hat{\beta}_1 = 0.48$, $\hat{K} = 86$, $\hat{\theta} = 0.70$, $\hat{\sigma}_e^2 = 0.036$, $\hat{\sigma}_d^2 = 0.33$. Bars show the observed distribution of population size during a 20-year period after first reaching carrying capacity. Soay Sheep on Hirta Island, with parameter estimates $\hat{\beta}_1 = 0.36$, $\hat{K} = 686$, $\hat{\theta} = 2.30$, $\hat{\sigma}_e^2 = 0.046$, $\hat{\sigma}_d^2 = 0.28$. Bars show the observed distribution of population size during a 42-year period with 5 years of missing data (Fig. 3.4).

for the mean time to extinction (Lande, 1993). The long-run growth rate of the population below K in the random catastrophe model is $\bar{r} = \ln \lambda - \varepsilon f$. To ensure eventual extinction in all the models, including those without demographic stochasticity, an absorbing boundary representing extinction was imposed at one individual instead of zero. This changes the limits of the integrals in formula 2.6b, with C and ∞ replaced respectively by 1 and K.

Asymptotic formulas for the mean time to extinction starting from K due to single stochastic factors in these simple models are summarized in Table 2.3 for populations with a large carrying capacity and a positive long-run growth rate below K. Under demographic stochasticity alone $T(K)$ increases almost exponentially with K, as previously established by several authors (MacArthur and Wilson, 1967; Richter-Dyn and Goel, 1972; Leigh, 1981; Goodman, 1987; Gabriel and Bürger, 1992). Under environmental stochasticity $T(K)$ increases as a power of K. The same qualitative scaling applies to random catastrophes, provided that more than a single catastrophe is required to extinguish the population starting from K (see Fig. 2.8). These results confirm those of Ludwig (1976), Leigh (1981), and Tier and Hanson (1981) for environmental stochasticity and Hanson and Tuckwell (1981) for random catastrophes.

For populations with a negative long-run growth rate, $T(K)$ increases only logarithmically with carrying capacity, regardless of the form of stochasticity. The logarithmic scaling applies even for deterministically declining populations, because if $N(t) = K e^{rt}$ with $r < 0$ the time to decline to a single individual is $-(\ln K)/r$. Logarithmic scaling of $T(K)$ as a function of K is therefore a general property of populations that are expected to decline. Large initial population size therefore does little to prolong the persistence of populations with a negative long-run growth rate.

Similarity of the scaling laws for mean time to extinction with increasing carrying capacity under environmental stochasticity and under random catastrophes justifies treating catastrophes as extremes of environmental stochasticity when investigating the mean time to extinction. For populations starting at K, demographic stochasticity

Table 2.3 Proportional asymptotic scaling of mean time to extinction, $T(K)$, as a function of carrying capacity, K, for different stochastic factors operating alone, with populations initially at K (Lande, 1993)

Demographic stochasticity[1]	$(1/K)e^{2\bar{\beta}K/\sigma_d^2}$ where $\bar{\beta} = \bar{\lambda} - 1$
Environmental stochasticity[1]	$K^{2\bar{r}/\sigma_e^2}$ where $\bar{r} = \bar{\beta} - \sigma_e^2/2$
Random catastrophes[1]	$K^{\zeta/\varepsilon}$ where $\zeta/(e^{\zeta} - 1) = f\varepsilon/\ln \lambda$
Expected decline[2]	$-(\ln K)/\bar{r}$

[1] Mean growth rate, $\bar{\beta}$, or long-run growth rate, $\bar{r}$, assumed positive.
[2] Long-run growth rate, $\bar{r}$, assumed negative under any stochastic factor.

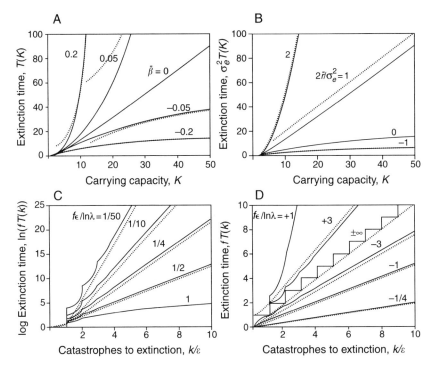

Fig. 2.8 Expected time to quasi-extinction (at $N = 1$), $T(K)$, starting from carrying capacity, K, as a function of K under single stochastic factors in a model of exponential growth to a ceiling at K. (A) Demographic stochasticity. (B) Environmental stochasticity. (C) and (D) Random catastrophes. Solid curves are from diffusion approximations in A and B, and from exact solution of delay differential equation using Laplace transformation in C and D. Dotted curves are asymptotic approximations with leading terms shown in Table 2.3. Parameters are defined in the text. (From Lande, 1993.)

poses a relatively weak risk of extinction in comparison with environmental stochasticity, except in populations with a small carrying capacity.

2.6.3 Demographic and environmental stochasticity

For most population models containing both demographic and environmental stochasticity the general expression for the mean time to extinction (eq. 2.6b) cannot be made more explicit because the integrals can not be evaluated analytically, but instead must be analyzed numerically. Informative results can nevertheless be obtained from the simple model of stochastic density-independent growth up to a ceiling at K. This model has infinitesimal mean and variance in equations 2.2a, with a reflecting boundary at K, and we impose an absorbing boundary or quasi-extinction threshold at $C < K$. Again for simplicity we suppose the population is initially at carrying capacity, $N_0 = K$. From equations 2.6a,b we find $s(N) = (N + \sigma_d^2/\sigma_e^2)^{-2\bar{\beta}/\sigma_e^2}$ where $\bar{\beta} = \bar{\lambda} - 1$ and

$$T(K) = \int_C^K \frac{1}{\bar{r}N} \left[\left(\frac{N + \sigma_d^2/\sigma_e^2}{C + \sigma_d^2/\sigma_e^2} \right)^{2\bar{r}/\sigma_e^2} - 1 \right] dN \qquad 2.8a$$

in which $\bar{r} = \bar{\beta} - \sigma_e^2/2$. This confirms that in the absence of demographic stochasticity, extinction never happens, as when C and σ_d^2 both equal zero we see that $T(K) = \infty$. Figure 2.9(A) shows that for C near zero, demographic stochasticity can greatly reduce the mean time to quasi-extinction.

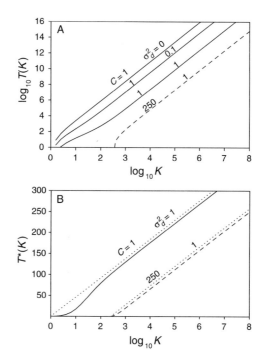

Fig. 2.9 Expected time to quasi-extinction $T(K)$ (panel A) and expect duration of the final decline $T^*(K)$ (panel B) starting from carrying capacity, K, for density-independent growth to a ceiling with both demographic and environmental stochasticity in density-independent growth rate (eqs 2.8a, 2.10). Note the different scales of the ordinates in the two panels, such that a straight line indicates a power function in panel A and a logarithmic function in panel B. Parameters are $\bar{\beta} = 0.07$, $\sigma_e^2 = 0.04$, with quasi-extinction threshold $C = 1$ for solid lines and $C = 250$ for dashed lines. Dotted lines in panel B represent the approximation $T^*(K) \lesssim [\ln(K/C)]/\bar{r}$.

Assuming a positive long-run growth rate, $\bar{r} > 0$, and that the carrying capacity is not very small, $K \gg \sigma_d^2/\sigma_e^2$, the integral can be accurately approximated as

$$T(K) \cong \frac{\sigma_e^2}{2\bar{r}^2} \left[\frac{K^{2\bar{r}/\sigma_e^2} - C^{2\bar{r}/\sigma_e^2}}{(\sigma_d^2/\sigma_e^2 + C)^{2\bar{r}/\sigma_e^2}} \right]. \qquad 2.8b$$

This has the same asymptotic scaling as under environmental stochasticity alone (Table 2.3), increasing as a power function K, and is virtually indistinguishable from the straight line portions of the graph in Figure 2.9(A).

With more general forms of density dependence the joint effects of demographic and environmental stochasticity on the mean time to extinction can be studied numerically. A logistic model with stochastic density-independent growth rate and stochastic carrying capacity was investigated (eq. 2.7). Mean times to extinction starting at the (expected) carrying capacity are depicted in Figure 2.10 as a function of carrying capacity for different values of the environmental variance in density-independent growth rate or in carrying capacity. This demonstrates that increasing environmental variance in either parameter generally decreases the mean time to extinction. Notice that rather large stochasticity in carrying capacity, as measured by the squared coefficient of variation $e^{\sigma_{\ln K}^2} - 1$, is necessary to decrease substantially the mean time to extinction.

Now consider the impact of environmental autocorrelation on extinction dynamics. Recall that short-term temporal autocorrelation in the environment can be incorporated in the diffusion approximation by magnifying the environmental variance by a factor equal to the total environmental autocorrelation summed or integrated over all time lags (Chapter 2.3.4). In conjunction with the finding that increasing environmental variance decreases the mean time to extinction, this implies that positive environmental autocorrelation increases extinction risk, whereas negative environmental autocorrelation decreases extinction risk. Computer simulations of populations with small density-independent growth rate corroborate this conclusion (Johst and Wissel, 1997). However, for discrete-time population models with high-density-independent growth rates, producing cyclic or chaotic dynamics that cannot be described by diffusion approximations, simulations have shown the opposite conclusion, that positive environmental autocorrelation decreases extinction risk (Ripa and Lundberg, 1996; Johst and Wissel, 1997).

This brings us to an important limitation of the models in this book, and of ecological models in general. With realistic population parameters, and assuming a stationary distribution of environments, populations with a carrying capacity sufficiently large to leave a good fossil record have expected persistence times, $T(K)$, much longer than a billion years (Figs 2.9A, 2.10). This greatly exceeds the mean duration of species observed in the fossil record (Table 2.1) for two reasons. First, our estimates of environmental stochasticity are likely too small due to undersampling based on limited duration of observations, generally no more than several decades, compared with the predicted time to extinction (Ludwig, 1996a, 1999; Fieberg and

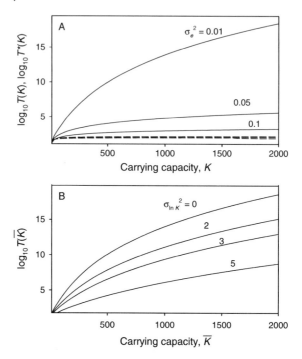

Fig. 2.10 (A) Expected time to extinction starting from carrying capacity, $T(K)$ (solid lines), and the expected duration of the final decline to extinction starting from carrying capacity $T^*(K)$, as functions of carrying capacity K (dashed lines) for values of environmental variance in density-independent growth rate σ_e^2 in the logistic model (eq. 2.7) with stochastic density independent growth rate and deterministic dependence ($\sigma_{\ln K}^2 = 0$). Other parameters are $\bar{\beta} = 0.1$ and $\sigma_d^2 = 1$. (B) Expected time to extinction starting from expected carrying capacity in the average environment, $T(\bar{K})$, as a function of $\bar{K}$ in the logistic model (eq. 2.7) with stochastic density-independent growth rate and stochastic carrying capacity for different values of $\sigma_{\ln K}^2$. Other parameters are $\bar{\beta} = 0.1$, $\sigma_e^2 = 0.01$, and $\sigma_d^2 = 1$.

Ellner, 2000). Second, real environments over very long times are actually non-stationary because of long-term climate change and extremely rare catastrophes, both of which are well documented in paleoclimatology (Roy *et al.*, 1996; Zachos *et al.*, 2001). Evolution of a species and its parasites, competitors, and predators also can produce nonstationary changes in population parameters. Nevertheless, simple stochastic population models with an observed or predicted distribution of environments are useful for describing relatively short-term processes, such as rapid local extinction and colonization in metapopulations (Chapter 4), the risk of extinction or quasi-extinction of small populations within a given time (Chapter 5), sustainable harvesting strategies (Chapter 6), and for analyzing data on stochastic community dynamics collected in a restricted time (Chapter 8).

2.7 Expected duration of the final decline

Most discussions of extinction dynamics concern the expected time to extinction or the probability of extinction within a given time (Chapter 5). Another useful description of extinction dynamics is the duration of the final decline to extinction or to a quasi-extinction threshold. This emphasizes how rapidly we might lose a population if conservation efforts cannot be instituted rapidly enough during the last phases of a species' existence, or how rapidly a population might become endangered.

Here we analyze the expected dynamics of the final decline to extinction, from when the population last crosses some specified value, N_0, at or below carrying capacity, K, and subsequently reaches a quasi-extinction threshold, $C < N_0$, without ever exceeding N_0. This can be described by a conditional diffusion process, excluding sample paths (or population trajectories) that cross above N_0. The infinitesimal mean and variance of the conditional process can be derived from those of the unconditional process (Karlin and Taylor, 1981),

$$M^*(N) = M(N) - \frac{s(N)V(N)}{S(N_0) - S(N)} \quad \text{and} \quad V^*(N) = V(N). \qquad 2.9a$$

The conditional process has the same infinitesimal variance as the unconditional process, whereas the infinitesimal mean is modified by an additional term that approaches $-\infty$ as N approaches N_0, which prevents the conditional trajectories from crossing N_0. These formulas facilitate the simulation of rare population trajectories leading directly from a large size to extinction despite a positive long-run growth rate below carrying capacity. Typical population trajectories for the conditional and unconditional processes starting from carrying capacity are illustrated in Figure 2.11.

From equations 2.6a and 2.9a it can be shown that the Green function for the conditional diffusion process is

$$G^*(N, N_0) = 2\frac{S(N_0) - S(N)}{s(N)V(N)}. \qquad 2.9b$$

The expected duration of the final decline from N_0 to C, conditional on the population not exceeding N_0 before hitting C, is then

$$T^*(N_0) = \int_C^{N_0} G^*(N, N_0)\, dN. \qquad 2.9c$$

For a population starting at carrying capacity, K, with a positive long-run growth rate below K, the expected duration of the final decline from K to extinction is generally much shorter than the mean persistence time of the population before extinction, $T^*(K) \ll T(K)$, as shown in Figure 2.10(A). Populations with a positive long-run growth rate below K are likely to fluctuate for a long time around K in a quasi-stationary distribution (provided K is not very small) crossing it many times

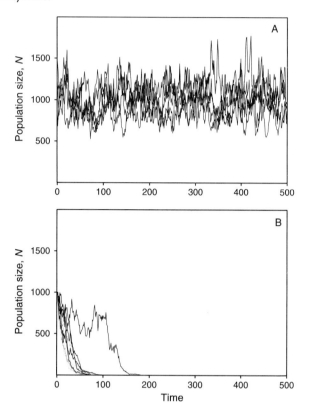

Fig. 2.11 Simulated population trajectories of unconditional (A) and conditional (B) processes starting at the carrying capacity $K = 1000$ in the logistic model with deterministic density dependence (eq. 2.7). Parameters are $\bar{\beta} = 0.01$, $K = 1000$, $\sigma_e^2 = 0.005$, and $\sigma_d^2 = 0.1$.

before the final crossing on the way to a relatively rapid final decline to extinction, as depicted by the simulated population trajectories in Figure 2.11. The relative rapidity of the final decline suggests that once the population first reaches K, the risk of extinction is nearly constant in time, so that after an initial time lag on the order of $T^*(K)$, the distribution of extinction times becomes approximately exponential (see below).

An intuitive idea of the expected duration of the final decline is provided by a reversibility property of conditional diffusion processes (Karlin and Taylor, 1981, pp. 387–8). Surprisingly, $T^*(K)$, the mean time of the final decline from K to C equals the mean time to grow from C to K conditional on not first passing below C. Thus populations with a high long-run growth rate below K, which can increase rapidly from small population size to K, are expected to have an equally short duration of the final decline to extinction. This occurs because starting from K a population with a high long-run growth rate below K can become extinct only through a chance series of unusually bad years in rapid succession, driving the population rapidly extinct before this can be prevented by its expected high rate of increase. However, unless K is very small, a high long-run growth rate below K also implies that extinction is a rare event and the population is likely to persist a long time before becoming rapidly extinct. For the stochastic logistic model (eq. 2.7), Figure 2.10(A) illustrates these points, also showing that $T^*(K)$ increases very slowly with K and is rather insensitive to environmental stochasticity.

This can be analyzed for the model of density-independent growth up to a ceiling or reflecting boundary at K, with a quasi-extinction threshold at $C < K$. The infinitesimal mean and variance in equation 2.2a again give $s(N)$ as above equation 2.8a, and we find from equations 2.9 that

$$T^*(K) = \int_C^K \frac{1}{\bar{r}N} \left[1 - \left(\frac{N + \sigma_d^2/\sigma_e^2}{K + \sigma_d^2/\sigma_e^2} \right)^{2\bar{r}/\sigma_e^2} \right] dN \qquad 2.10$$

where $\bar{r} = \bar{\beta} - \sigma_e^2/2$. Assuming a positive long-run growth rate, $\bar{r} > 0$, this formula reveals that $T^*(K) \lesssim [\ln(K/C)]/\bar{r}$. Thus the expected duration of the final decline to quasi-extinction is less than the time for deterministic exponential growth from C to K at the long-run growth rate. This simple upper bound provides a fairly accurate asymptotic approximation to $T^*(K)$ (see Fig. 2.9B), showing that the expected duration of the final decline to quasi-extinction increases logarithmically with increasing carrying capacity and is inversely proportional to the long-run growth rate.

2.8 Distribution of time to extinction

In conservation biology the quantitative assessment of extinction risk is based on estimating the cumulative probability of population extinction within a specified time (Shaffer, 1981; Mace and Lande, 1991; IUCN, 2001). This requires estimating the distribution of time to extinction and its sum or integral up to the given time. Also of practical interest is the risk of quasi-extinction (Ginzburg *et al.*, 1982; Burgman *et al.*, 1993), the probability of population decline to a certain small size within a given time. Analytical formulas for the distribution of time to extinction or quasi-extinction (eq. 2.5a) exist in only a few cases. Usually it is necessary to simulate the distribution up to a specified time, as illustrated below and in Chapter 5.

2.8.1 Quasi-extinction in a density-independent population

We assume an initial population size N_0 well below carrying capacity so that density dependence can be ignored, and impose an absorbing boundary or quasi-extinction threshold at C large enough to neglect demographic stochasticity and Allee effects. On a log scale, $X = \ln N$, the diffusion approximation has infinitesimal mean and variance $M(X) = \bar{r}$ and $V(X) = \sigma_r^2$ (see above eq. 2.2b), and we denote $X_0 = \ln N_0$

and $c = \ln C$. This represents Brownian motion with drift and an absorbing barrier. The probability distribution of the time, t, until quasi-extinction is given by the inverse Gaussian distribution (Cox and Miller, 1965; Johnson and Kotz, 1970; Lande and Orzack, 1988),

$$\Xi(X_0, t) = \frac{X_0 - c}{\sqrt{2\pi \sigma_r^2 t^3}} \exp\left\{\frac{-(X_0 - c + \bar{r}t)^2}{2\sigma_r^2 t}\right\} \qquad 2.11$$

The integral of this distribution from time 0 to t gives the cumulative probability of quasi-extinction before time t for any value of the long-run growth rate. When $\bar{r} < 0$ eventual quasi-extinction is certain, because as $t \to \infty$ the cumulative probability approaches unity. However, when $\bar{r} > 0$ eventual quasi-extinction is not certain but occurs with probability $e^{-2(X_0 - c)\bar{r}/\sigma_r^2}$. With a large carrying capacity this can be interpreted as the probability of reaching carrying capacity before quasi-extinction.

For $\bar{r} > 0$ a proper probability distribution of time to quasi-extinction does not exist because quasi-extinction is not certain. However, considering the conditional diffusion process, including only those sample paths that reach c (before reaching ∞), a proper conditional distribution of quasi-extinction time can be derived by dividing the unconditional distribution by the probability of eventual quasi-extinction, $\Xi^*(X_0, t) = \Xi(X_0, t)e^{2(X_0 - c)\bar{r}/\sigma_r^2}$. Remarkably this is exactly the same distribution as in equation 2.11 but with $\bar{r}$ replaced by $|\bar{r}|$ (Lande and Orzack, 1988). The mean time to quasi-extinction, conditional on the event and regardless of the sign of $\bar{r}$ is then $T(X_0) = (X_0 - c)/|\bar{r}|$, which is the same for a population undergoing a deterministic exponential decline, depending only on the magnitude and not the sign of the long-run growth rate and independent of the environmental variance. The variance of the time to quasi-extinction is $3(X_0 - c)\sigma_e^2/|\bar{r}|^3$. Note that, even with a negative long-run growth rate, the distribution of time to quasi-extinction has a strong positive skew, with an initial period of low extinction risk, as depicted in Figure 2.12.

2.8.2 Extinction in a density-dependent population

Now consider extinction in a density-dependent population with demographic and environmental stochasticity. Assuming a positive long-run growth rate below K, the distribution of time to extinction is asymptotically exponential, as shown analytically (Nobile *et al.*, 1985) and by simulation (Goodman, 1987). Analysis of the unconditional and conditional mean times to extinction (after eq. 2.9c), suggested that for N_0 near K there is an initial period on the order of $T^*(K)$ years before the asymptotic exponential decay in the distribution of time to extinction. Thus, following a brief period required to approach the quasi-stationary distribution of population size, the distribution of time to extinction is approximately exponential, implying a nearly constant risk of extinction per unit time, as shown in Figure 2.13. If the population is initially near extinction, then the extinction risk per unit time may be elevated during the initial phase before approaching a constant (Fig. 2.13; Ludwig, 1996*a*; Grimm and Wissel, 2003).

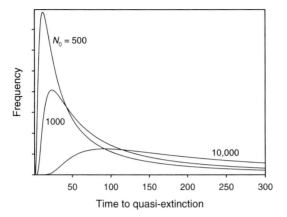

Fig. 2.12 Distribution of the time to quasi-extinction under density-independent growth (eq. 2.11) for initial population sizes $N_0 = 500$, 1000, and 10,000 and other parameters $\bar{r} = -0.03$, $\sigma_e^2 = 0.075$. Absorbing boundary at population size $C = 100$ is assumed large enough to neglect demographic stochasticity and Allee effects.

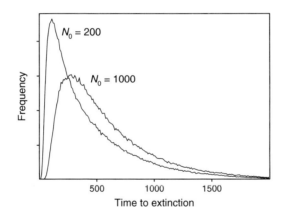

Fig. 2.13 Simulated distribution of time to extinction in the logistic model (eq. 2.7) with deterministic density dependence for different initial population sizes N_0. Number of simulations was 10^7 with parameters $\bar{\beta} = 0.01$, $K = 1000$, $\sigma_e^2 = 0.02$, $\sigma_d^2 = 1$.

2.9 Summary

The mean durations of species from various animal taxa in the fossil record are a few to several million years. Most paleo-extinctions occurred as apparently random background extinctions, whereas modern anthropogenic extinctions constitute the beginning of the sixth great mass extinction in the history of life on earth. In documented extinctions of modern vertebrate species, population trajectories usually have occurred by irregular declines over many years, with the final extinction happening from small population size. Diffusion approximation provides a powerful method for analyzing stochastic dynamics and extinction in simple population models with no age structure and small expected rates of population growth. Analysis reveals that demographic stochasticity can create a stochastically unstable equilibrium at a small population size, below which most population trajectories tend to decrease, resulting in rapid extinction, similar to an Allee effect.

Under density-dependent population growth with a stationary distribution of environmental effects, for a population initially at or near carrying capacity the mean time to extinction increases as a power of the carrying capacity, and after a short time with low extinction risk the distribution of time to extinction is approximately exponential. With realistic parameters for expected population growth rate and stochasticity, for populations with moderate or large carrying capacity the models predict mean times to extinction far longer than those observed in the fossil record. This occurs because short-term studies underestimate environmental stochasticity, and because long-term environmental changes are not stationary. However, the expected duration of the final decline from carrying capacity to extinction usually is relatively short. Simple population models with observed or predicted distributions of environmental effects are useful for analyzing short-term processes, such as the extinction of small or local populations.

3

Age structure

3.1 Deterministic density-independent dynamics

Life history and age structure affect the dynamics of populations when individuals reproduce in multiple years, with survival and reproductive rates that change with age. Such an iteroparous life history commonly occurs in higher animals and plants, creating multiple time delays in the population dynamics, affecting the stability and fluctuations of population size and age structure. Classical demographic theory of age-structured populations describes the deterministic dynamics under density-independent growth (Euler, 1760; Lotka, 1924; Leslie, 1945, 1948). Assuming a constant environment, with population size large enough to neglect demographic stochasticity, the classical theory can be used to predict changes in population size and age structure in two situations. The first is obviously for density-independent populations well below carrying capacity. The second is for density-dependent populations over short timespans during which the population density does not change much. The classical theory has proven especially useful to describe and predict changes in human populations over a few decades (Keyfitz, 1968). It is also routinely applied to wild species to calculate current population growth rates under prevailing environmental conditions, based on estimates of age-specific survival and reproduction (Caughley, 1977; Carey, 1993).

Accurate determinations of individual age are difficult or impossible in many species, unless individuals have been identified and followed from birth. Furthermore, in contrast to age-structured models, stage-structured models often require far fewer parameters to describe the life history and are therefore likely to be statistically more reliable. This motivated the development of theories describing the dynamics of populations structured by individual size or developmental stage (Lefkovitch, 1965; Caswell, 2001). However, several aspects of the theory are more transparent and explicit for age structure than for stage structure. We use both approaches below in explicating deterministic and stochastic dynamics of age-structured populations.

3.1.1 Classical age-structured life history

Deterministic demography classically concerns only the female population, assuming that all females are fertilized regardless of population size. Following Leslie (1966), Mertz (1971), and Caswell (2001) we consider a population with synchronized annual breeding and annual censuses conducted immediately prior to reproduction. The number of individuals of age i in year t is denoted as $n_i(t)$ for $i = 1, 2, \ldots, \omega$ where

ω is the maximum length of life. The life history of the population is described by the age-specific vital rates, s_i, the probability of individual survival from age i to $i + 1$, and f_i, the product of fecundity for females of age i multiplied by the probability of offspring survival during their first year of life (counting only female offspring per female per year). We assume that the vital rates are constant in time and that population size is large enough to neglect demographic stochasticity. The projection equations are

$$n_1(t + 1) = \sum_{i=1}^{\omega} f_i n_i(t) \qquad\qquad 3.1a$$

$$n_{i+1}(t + 1) = s_i n_i(t) \quad \text{for } i = 1, 2, \dots, \omega - 1. \qquad 3.1b$$

For iteroparous species, in which the reproductive rates of at least two successive ages are nonzero, populations asymptotically approach a stable age distribution with geometric growth in time. This can be proven by repeatedly substituting equation 3.1b into 3.1a to recast the projection equations in the form of a renewal equation for the first age class,

$$n_1(t) = \sum_{\tau=1}^{\omega} l_\tau f_\tau n_1(t - \tau) \quad \text{with } l_\tau = \prod_{i=1}^{\tau-1} s_i \text{ for } \tau = 2, \dots, \omega \qquad 3.1c$$

where l_τ is the survival function giving the probability of survival from age 1 to age τ with $l_1 = 1$. Population dynamics described by stage-structured life histories can be similarly recast as an equation in a single variable with time delays (reviewed by Nisbet, 1997). Assuming an asymptotic solution for geometric growth at the multiplicative rate λ per year, $n_1(t) = \lambda^\tau n_1(t - \tau)$, using this to substitute for $n_1(t - \tau)$ in equation 3.1c, and cancelling the common factor of $n_1(t)$ from both sides produces the Euler–Lotka equation,

$$\sum_{\tau=1}^{\omega} l_\tau f_\tau \lambda^{-\tau} = 1. \qquad\qquad 3.2$$

This polynomial equation has ω solutions or roots that may be real or complex and can be obtained numerically. The unique real positive solution λ_1 has the largest modulus (the vector length of a complex number). The real part of the root with the second largest modulus, $\text{Re}(\lambda_2)$, determines the multiplicative rate of approach to the stable age distribution, $\text{Re}(\lambda_2)/\lambda_1$. For simplicity, the dominant eigenvalue λ_1 will subsequently be denoted simply as λ.

When the stable age distribution has been achieved, the total population size, N, or any linear combination of age classes, grows geometrically in time with a multiplicative growth rate λ. The stable age distribution can be derived by repeatedly substituting equation 3.1b into itself using the geometric growth formula $n_i(t + 1) = \lambda n_i(t)$ to obtain

$$u_i = l_i \lambda^{-i} \Bigg/ \sum_{\tau=1}^{\omega} l_\tau \lambda^{-\tau} \qquad\qquad 3.3$$

where the denominator is a normalization constant to ensure that $\sum_{i=1}^{\omega} u_i = 1$. This reveals that the stable age distribution is proportional to the survival function, l_i, but is also affected by the fecundity schedule, f_i, through the multiplicative growth rate, λ. A stable population with $\lambda = 1$ has a stable age distribution simply proportional to the survival function, l_i. Comparing populations with the same survival function, in a growing population with high fecundity and $\lambda > 1$ the stable age distribution is weighted toward young individuals, whereas in a declining population with low fecundity and $\lambda < 1$ the stable age distribution is weighted toward old individuals.

The generation time is defined as the mean age of mothers of newborn individuals when the population has achieved a stable age distribution. In view of formula 3.2 this is

$$T = \sum_{\tau=1}^{\omega} \tau l_\tau f_\tau \lambda^{-\tau}. \qquad 3.4$$

There are other possible definitions of generation time (Charlesworth, 1994; Caswell, 2001) but this definition naturally appears in a variety of contexts, as shown below. Like the stable age distribution, the generation time depends on λ. More rapidly growing populations have shorter generation times. For realistic life histories in iteroparous species, populations generally approach the stable age distribution and the asymptotic multiplicative growth rate λ within a few generations, as exemplified in Figure 3.1.

The reproductive value of a female of age i describes the contribution of females of different ages to the future growth of the population (Fisher, 1930; Leslie, 1948). For the discrete-time model analyzed here the reproductive value is (Charlesworth, 1994)

$$v_i = \frac{\lambda^i}{l_i} \sum_{\tau=i}^{\omega} l_\tau f_\tau \lambda^{-\tau}. \qquad 3.5a$$

In most species the juvenile survival rates, especially in the first year, are substantially lower than adult survival rates, until old age when senescent declines in survival and reproduction may occur (Deevey, 1947; Gaillard *et al.*, 1994; Nichols *et al.*, 1997; Loison *et al.*, 1999). For such species the reproductive value generally increases until the age of first reproduction and then declines with age.

3.1.2 Matrix formulation

The classical theory of age- or stage-structured population dynamics also can be formulated in vector–matrix notation (Leslie, 1945, 1948; Caswell, 2001), which facilitates the derivation of certain results. Writing the state vector of ages or stages in the population as a column vector $\mathbf{n}(t) = (n_1(t), \ldots, n_\omega(t))^{\mathsf{T}}$ where the superscript ^{T} denotes matrix transposition, the projection equations 3.1 are simply $\mathbf{n}(t+1) = \mathbf{A}\mathbf{n}(t)$ where the square projection matrix $\mathbf{A}$ has in the ith row and jth column the element $A_{i,j}$ giving the annual rate of individual transition from state j to state i. For an age-structured population $\mathbf{A}$ is known as the Leslie matrix, which has all elements zero except for the annual fecundities in the top row $A_{1,j} = f_j$ and annual survival rates in

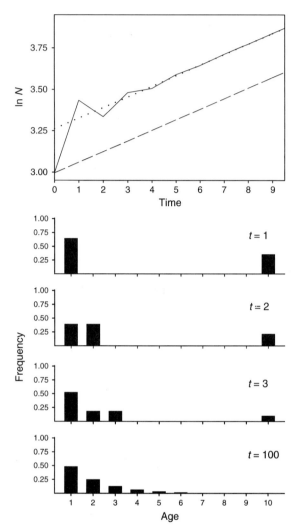

Fig. 3.1 Deterministic population growth on log scale (upper panel) and change with time in age distribution (lower panels) in a Lefkovitch matrix model with 10 stage classes, corresponding to ages 1–9 and the last stage class contains individuals aged 10 and older. Annual survival of each class is 0.55. Annual fecundity times first year survival is $f_1 = 0$ for the first age class (aged 1) and all subsequent classes have annual fecundity times first year survival of $f_i = 1$ for $i \geq 2$. Initial condition at $t = 0$ was 20 individuals in the oldest stage class. Solid line describes the growth of the total population size, $N(t)$. Dashed line is the prediction assuming the population is always in a stable age distribution, $\ln N_0 + \lambda t$ where $\lambda = 1.066$. Dotted line describes the population growth accounting for the initial age distribution using reproductive value (eq. 3.5b).

the subdiagonal $A_{j+1,j} = s_j$. For a stage-structured population the projection matrix is called the Lefkovitch matrix, which can have other nonzero elements (Lefkovitch, 1965; Caswell, 2001). The asymptotic multiplicative growth rate of the population, λ, is the dominant eigenvalue of the projection matrix **A**. The stable age distribution, denoted by the column vector $\mathbf{u} = (u_1, \ldots, u_\omega)^\mathsf{T}$ and reproductive values denoted by the row vector $\mathbf{v} = (v_1, \ldots, v_\omega)$, apart from an arbitrary multiplicative constant, are the corresponding right and left dominant eigenvectors of the projection matrix, $\mathbf{Au} = \lambda\mathbf{u}$ and $\mathbf{vA} = \lambda\mathbf{v}$. The Perron–Frobenius Theorem for matrices with non-negative elements guarantees that the dominant eigenvalue is unique, real, and positive, and the elements of **u** and **v** are non-negative.

Because the total population size only asymptotically approaches the constant multiplicative growth rate λ, a naive projection of the initial population size at the asymptotic rate, $\lambda^t N(0)$, would eventually result in a constant proportional deviation from the actual population size. In contrast, it is remarkable that the total reproductive value in the population $\mathbf{vn}(t)$ always grows at the multiplicative rate λ, even when the initial age distribution departs from the stable age distribution (Fisher, 1930; Leslie, 1948; Charlesworth, 1994), which follows directly from the projection formula and eigenvalue equation, $\mathbf{vn}(t + 1) = \mathbf{vAn}(t) = \lambda\mathbf{vn}(t)$. This allows a correction to be made to the initial age distribution using reproductive value weighting to achieve an accurate asymptotic projection of the total population size, as follows. Projecting the initial age distribution, $\mathbf{n}(t) = \mathbf{A}^t\mathbf{n}(0)$, a well-known result in matrix theory (from the spectral decomposition of a matrix) is that as $t \to \infty$ the matrix power approaches the asymptotic value $\mathbf{A}^t \approx \lambda^t\mathbf{uv}/(\mathbf{vu})$ (Charlesworth, 1994; Caswell, 2001) where the 'inner product' **vu** is a scalar normalization constant and the 'outer product' **uv** is a matrix with elements $u_i v_j$. Substituting this into the projected age distribution, multiplying on the left by the row vector $(1, \ldots, 1)$ to sum the vector elements, and using $\sum n_i(t) = N(t)$ and $\sum u_i = 1$, gives

$$N(t) \approx \lambda^t \frac{\mathbf{vn}(0)}{\mathbf{vu}} \qquad\qquad 3.5\mathrm{b}$$

as illustrated in Figure 3.1.

The sensitivity of λ with respect to a small change in one of the nonzero vital rates, $A_{i,j}$, holding the other vital rates constant, can be expressed in terms of elements of the right and left dominant eigenvectors, $\partial\lambda/\partial A_{i,j} = v_i u_j/(\mathbf{vu})$ (Caswell, 1978, 2001, p. 209). For an age-structured population the sensitivity coefficients can be derived most simply by implicit differentiation of the Euler–Lotka equation 3.2, treating λ as a function of the vital rates (Hamilton, 1966; Demetrius, 1969; Charlesworth, 1994),

$$\frac{\partial\lambda}{\partial f_i} = \frac{\lambda^{-i+1}l_i}{T} \quad \text{and} \quad \frac{\partial\lambda}{\partial s_i} = \frac{\lambda^{-i}l_i v_{i+1}}{T}. \qquad\qquad 3.6$$

Sensitivity coefficients are useful in population management and conservation to suggest where in the life history management interventions can produce the largest increases in population growth rate (Heppell *et al.*, 1994; Ehrlén and van Groenendael, 1998). They appear in approximations for the standard error of an estimate of λ in a constant environment (eq. 3.7) and the long-run growth rate of a population in a fluctuating environment (eq. 3.9). Sensitivity coefficients also describe the intensity of natural selection on the vital rates (Hamilton, 1966; Lande, 1982a,b; Charlesworth, 1994; Caswell, 2001). The sensitivity of $\ln \lambda$ with respect to the natural log of a quantity is known as an elasticity coefficient (de Kroon et al., 1986; Caswell, 2001). The elasticity of λ with respect to a (life history) parameter p is denoted as

$$e_p = \frac{\partial \ln \lambda}{\partial \ln p} = \frac{p}{\lambda} \frac{\partial \lambda}{\partial p}.$$

3.1.3 Estimating population growth rate

An estimate of the multiplicative growth rate of a population in a constant environment, $\hat{\lambda}$, can be obtained by substituting estimates of the vital rates into the Euler–Lotka equation 3.2 and solving numerically for $\hat{\lambda}$. To simplify notation the age-specific survival and reproductive rates are collectively denoted by a single subscripted symbol π_i for $i = 1, \ldots, 2\omega - 1$. The accuracy of an estimate, $\hat{\lambda}$, described by its standard error, the square root of the sampling variance, Var[$\hat{\lambda}$], can be obtained by two basic methods. The first is bootstrapping of data on the vital rates, which involves empirical estimation of the sampling distribution of $\hat{\lambda}$ by repeated resampling of the data on individual fitness (Efron and Tibshirani, 1993), and the second is from the approximate equation (Lande, 1988; Alvarez-Buylla and Slatkin, 1993; Caswell, 2001)

$$\text{Var}[\hat{\lambda}] = \sum_i \sum_j \left(\frac{\partial \lambda}{\partial \pi_i} \right) \left(\frac{\partial \lambda}{\partial \pi_j} \right) \text{Cov}[\hat{\pi}_i, \hat{\pi}_j] \qquad 3.7$$

derived by taking the variance of a first-order Taylor expansion of λ as a function of the vital rates. Notice in this formula the appearance of sensitivity coefficients (eq. 3.6). With sufficient data the estimates of λ will be approximately normally distributed with 95% confidence interval 1.96 times the standard error on either side of λ. If different vital rates are estimated from independent data, then the sampling covariances between them vanish, and formula 3.7 contains only the sampling variance of each vital rate times its squared sensitivity coefficient. In a sample of size N_i an estimated survivorship has a binomial distribution with Var[$\hat{\pi}_i$] $= \hat{\pi}_i(1 - \hat{\pi}_i)/N_i$ and an estimated reproductive rate has a sampling variance $\sigma_{\pi_i}^2/N_i$. Monte Carlo simulation indicates that formula 3.7 gives reliable estimates of the standard error, similar to those obtained by bootstrapping, provided that the coefficients of variation of the vital rates are less than about 0.5 (Alvarez-Buylla and Slatkin, 1993).

3.1.4 A simple stage-structured life history

Many avian and mammalian species have annual survival and reproductive rates that are nearly constant and independent of age; in such species the great majority of individuals die before reaching the age of senescence, which is therefore of little demographic consequence (Deevey, 1947; Gaillard *et al.*, 1994; Nichols *et al.*, 1997; Loison *et al.*, 1999). Denoting the age of first reproduction as α, and the adult annual survival and reproductive rates as s and f, we have $l_\tau = l_\alpha s^{\tau-\alpha}$ and $f_\tau = f$ for $\tau \geq \alpha$ with $f_\tau = 0$ for $\tau < \alpha$. The Euler–Lotka equation and the generation time formula (eqs 3.2 and 3.4) can then be summed using $\omega = \infty$ to give (Lande, 1988)

$$l_\alpha f \lambda^{-\alpha} + s\lambda^{-1} = 1 \quad \text{and} \quad T = \alpha + s/(\lambda - s). \qquad 3.8$$

This and related stage-structured models have been employed by several authors for a variety of demographic and evolutionary studies (Caswell, 2001, p. 192). For example, this model was applied to data on the Northern Spotted Owl (*Strix occidentalis caurina*) to estimate λ and its standard error obtained from formula 3.7 using appropriate sensitivity coefficients. The initial estimate with its standard error $\hat{\lambda} = 0.961 \pm 0.029$ was not significantly different from $\lambda = 1$ (Lande, 1988). A similar model and more extensive data from 11 study sites, with survival rates estimated from mark–recapture methods and reproductive rates estimated from field surveys, later showed that λ was significantly less than 1 at all but one of the sites (Forsman *et al.*, 1996).

3.2 Long-run growth rate in a random environment

Environmental fluctuations cause changes not only in population growth rate, but also in the age-specific vital rates and hence in age structure, according to the time-dependent matrix projection $\mathbf{n}(t + 1) = \mathbf{A}(t)\mathbf{n}(t)$. In this section we consider density-independent growth of an age-structured population in a random environment, where each of the vital rates may fluctuate in time with a stationary distribution, and different vital rates also may be correlated. Positive correlations between vital rates arise because environments that affect reproduction or survival in one age class have similar effects on other age classes. Negative correlations between vital rates can arise from life-history trade-offs, for example between adult reproduction and survival, or between early and late reproduction (Lande, 1982*a,b*; Charlesworth, 1994). For simplicity we assume no environmental autocorrelation.

Let $N(t)$ be the total size of an age-structured population at time t. For density-independent growth in a random environment, neglecting demographic stochasticity, the long-run growth rate can be defined as the asymptotic change per unit time in log population size, and the long-run variance in growth rate can be similarly defined,

$$\bar{r} = \lim_{t \to \infty} \frac{1}{t} E\left[\ln \frac{N(t)}{N(0)}\right] \quad \text{and} \quad \sigma_r^2 = \lim_{t \to \infty} \frac{1}{t} \text{Var}\left[\ln \frac{N(t)}{N(0)}\right]. \qquad 3.9a$$

As for populations without age structure, the long-run growth rate of an age-structured population determines whether the total population (or any linear combination of age classes) tends to increase, $\bar{r} > 0$, or decrease, $\bar{r} < 0$, and $\ln N(t)$ is asymptotically normally distributed (Cohen, 1977, 1979).

An approximation for the long-run growth rate of an age-structured population with small environmental stochasticity was derived by Tuljapurkar (1982), generalizing equation 1.4,

$$\bar{r} = \ln \lambda - \frac{\sigma_r^2}{2} \quad \text{where } \sigma_r^2 \cong \frac{1}{\lambda^2} \sum_i \sum_j \left(\frac{\partial \lambda}{\partial \pi_i} \right) \left(\frac{\partial \lambda}{\partial \pi_j} \right) \text{Cov}[\pi_i, \pi_j]. \qquad 3.9b$$

Here λ is the asymptotic multiplicative growth rate in the average environment, which is actually same as the asymptotic multiplicative growth rate of the total population size in a fluctuating environment. This can be shown from the matrix projection equation in a fluctuating environment, $\mathbf{n}(t + 1) = \mathbf{A}(t)\mathbf{n}(t)$, by taking expectations using the assumption of independent environments, giving $\bar{\mathbf{n}}(t + 1) = \bar{\mathbf{A}}\bar{\mathbf{n}}(t)$ where $\bar{\mathbf{A}}$ is the average projection matrix with elements $\bar{A}_{ij}$. Thus by a formula analogous to 3.5b the expected total population size, $\bar{N}(t)$, grows asymptotically at the multiplicative rate λ, the dominant eigenvalue of $\bar{\mathbf{A}}$, the solution of the Euler–Lotka equation in the average environment. The long-run variance in growth rate, σ_r^2, involves a formula similar to the sampling variance of λ (eq. 3.7) but here the sensitivity coefficient of λ with respect to the vital rate π_i is evaluated in the average environment, and $\text{Cov}[\pi_i, \pi_j]$ is the environmental covariance between the vital rates.

Because σ_r^2 is a variance that cannot be negative, formula 3.9b demonstrates that environmental stochasticity decreases the long-run growth rate of an age-structured population, as depicted in Figure 3.2. Furthermore, $\bar{r}$ is less than the asymptotic growth rate of $\ln \bar{N}(t)$, as established in Chapter 1 for a population without age structure. In contrast to populations without age structure (eqs 1.8), in an age-structured population environmental autocorrelation does affect the long-run growth rate (Tuljapurkar, 1982, 1990; Caswell, 2001).

Tuljapurkar's approximation has good accuracy for small or moderate fluctuations in the vital rates, with coefficients of variation in the vital rates up to at least 30%, as shown in Table 1.3 for populations without age structure. With age structure, accuracy is preserved for even larger coefficients of variation in vital rates that have small sensitivity coefficients, or for vital rates with negative covariance (Lande and Orzack, 1988; Orzack, 1997), as illustrated in Figure 3.3. Estimates of temporal fluctuations in vital rates (including changes due to both environmental stochasticity and density dependence) demonstrate that the vital rates with higher environmental variances also have smaller sensitivity coefficients (Pfister, 1998; Sæther and Bakke, 2000), implying that equation 3.9b is likely to be accurate in many cases.

Two naive approaches to calculation of the long-run growth rate are rather inaccurate because they fail to account for fluctuations in the age structure. The first is to define incorrectly the long-run growth rate as the product of values of λ for the different environments (Cohen, 1979; Tuljapurkar, 1989; Caswell, 2001, p. 401). The second is taking the expectation of a second-order Taylor expansion of $r = \ln \lambda$ as a function of the vital rates, as in the derivation of equation 1.4, which produces a formula like 3.9 but with $\partial^2 \lambda / \partial \pi_i \partial \pi_j$ instead of the product of sensitivity coefficients. A third approach is less accurate than Tuljapurkar's approximation for a subtler reason. Because $\ln N(t)$ is asymptotically normal, it might be thought that the transformation formulas relating the mean and variance of a lognormal distribution to the mean and variance of the corresponding normal distribution on the log scale (as in eqs 1.5a) could be used to obtain an exact expression for the long-run growth rate. However, despite the asymptotic normality of $\ln N(t)$, the distribution of $N(t)$ is not asymptotically lognormal to the same degree of accuracy because the deviation of $\ln N(t)$ from normality becomes magnified by the exponential transformation.

Although the long-run growth rate of an age-structured population is defined asymptotically for density-independent growth, the simulations in Figure 3.2 show that it gives an accurate indication of the average rate of increase of the log of total population size even over a few generations, for a population near stable age distribution, similar to using λ for an age-structured population in a constant environment (Fig. 3.1). The long-run growth rate is therefore a useful concept for understanding the stochastic growth of real age-structured populations, even with density dependence, provided that the population density does not change very rapidly.

For example, like many species with long generations the Northern Spotted Owl has a high adult annual survival rate that fluctuates little with time, and substantial

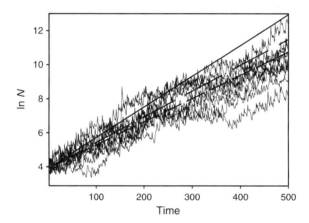

Fig. 3.2 Population trajectories from a age-structured model in a fluctuating environment with no demographic stochasticity and no density dependence. Life history is the same as that in Fig. 3.1 except that the expected annual fecundity times first year survival of the oldest stage class (age ≥ 10) is $f_{10} = 1.5$, and annual survival rates are $s_1 = 0.55$ in the second year, $s_2 = 0.7$ in the third year, and $s_i = 0.85$ for $i \geq 3$. Initial population size is 50 distributed according to the stable age distribution in the average environment. The environmental variance in adult annual fecundity is 3.0 with the same value for all adult age classes in a given year. Solid line gives the deterministic prediction assuming a stable age distribution in a constant (average) environment. Long-dashed line is the expected long-term growth according to Tuljapurkar's first-order approximation (eq. 3.9b). Short-dashed line gives the expected long-term growth obtained from simulations.

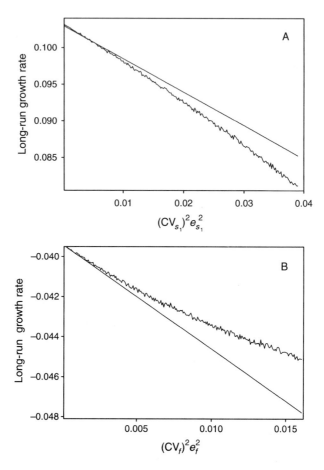

Fig. 3.3 Tuljapurkar's first-order approximation to the long-run growth rate (solid line) in relation to the expected long-run growth rate obtained from simulations for different levels of environmental stochasticity in (A) second year survival s_1 or (B) annual adult fecundity times first year survival, f. Number of simulations was 100,000. Life history is similar to that in Figure 3.1. Parameters in (A) are $f_1 = 0$ and $f_i = 1$ for stage classes $2 \leq i \leq 10$, with $s_1 = 0.35, s_2 = 0.65$, $s_i = 0.85$ for $3 \leq i \leq 10$. Second year survival s_1 takes two different values at random with equal probability among years. Parameters in (B) are $f_1 = 0$ and $f_i = 1.5$ for stage classes $2 \leq i \leq 10$, with $s_1 = 0.25, s_2 = 0.65, s_i = 0.85$ for $3 \leq i \leq 10$. For stage classes $i \geq 2$ annual fecundity is lognormally distributed among years with a constant mean and the same value for all adult age classes a given year. The level of environmental variation on the horizontal axis is scaled in terms of the coefficient of variation and the sensitivity coefficient for the variable life history parameters.

temporal variation in recruitment of yearlings. Using upper bounds for the environmental variance and covariance of these vital rates gave $\exp\{\bar{r}\} \cong 0.98\lambda$ (Lande, 1988). Extensive data on a population of Red Deer (*Cervus elaphus*) on the Isle of Rum revealed a qualitatively similar life history, with high adult survival rates and a substantially larger environmental variance in fecundity than in survival. Analysis of data from 1982 to 1991, beginning 10 years after cessation of hunting, yielded $\exp\{\bar{r}\} \cong (0.998)\lambda$ (Benton *et al.*, 1995). Despite a large environmental variance in annual fecundity, for such species environmental stochasticity produces only a small reduction in the long-run growth rate because of the small sensitivity of λ with respect to annual fecundity. In other words, even in populations with a very large coefficient of variation in recruitment of yearlings, with occasional good years and frequent bad years for fecundity times first year survival, a long adult life expectancy tends to average out the good and bad years, reducing the impact of environmental stochasticity on the long-run growth rate (Lande, 1988; Pimm, 1991; Benton *et al.*, 1995; Sæther *et al.*, 2002*b*).

The effect of environmental stochasticity on the long-run growth rate is larger for species with shorter generations, because the environmental variance and hence the long-run variance in growth rate tends to be larger (see Tables 1.2 and 1.4), or in other words because good and bad years are much less subject to averaging within individual lifetimes. This is illustrated by statistical analysis and demographic modeling of populations of a relatively short-lived species, the Grey Partridge, which displayed a rather substantial influence of environmental stochasticity on the long-run growth rate, with the two largest populations having $\exp\{\bar{r}\} \cong (0.62)\lambda$ and $\exp\{\bar{r}\} \cong (0.66)\lambda$ (Bro *et al.*, 2000).

A diffusion approximation for the natural logarithm of total population size, $X = \ln N$, can be derived for density-independent growth of an age-structured population in a random environment (Lande and Orzack, 1988). Neglecting demographic stochasticity, the infinitesimal mean and variance are, respectively, the long-run growth rate and the long-run variance in growth rate, $\bar{r}$ and σ_r^2, which are constants, representing Brownian motion with drift, as in the corresponding model without age structure (above eq. 2.2b). The accuracy of the diffusion approximation is improved by employing reproductive value weighting of the initial age distribution, as in the deterministic model 3.5b, with reproductive values calculated for the population in the average environment, $X(0) = \ln(\mathbf{vn}(0)/(\mathbf{vu}))$. Dennis *et al.* (1991) and Gaston and Nicholls (1995) estimated the infinitesimal mean and variance for this diffusion approximation from time series observations on total population size to evaluate the risk of population decline to quasi-extinction for several threatened and endangered species, assuming density-independent growth and neglecting demographic stochasticity. Holmes (2001, 2002) and Holmes and Fagan (2002) extended this approach to deal with inaccurately estimated population time series.

This diffusion approximation for the dynamics of an age-structured population can be extended to include demographic stochasticity, following Engen *et al.* (2003*b*). Define the ith vital rate of an individual $w_i = \mu_i + \delta_i$ to be the sum of an environmental component identical for all individuals, μ_i, and an independent demographic component, δ_i, with $E[\delta_i] = 0$ (similar to eq. 1.1), and let n_i be the number of individuals in the age class corresponding to the ith vital rate. Averaging over all individuals within each class and defining $\pi_i = \bar{w}_i$ gives the covariances of the mean vital rates in the population (generalizing eq. 1.3)

$$\text{Cov}[\pi_i, \pi_j] = \text{Cov}[\mu_i, \mu_j] + \delta_{ij} \frac{\text{Var}[\delta_i]}{n_i}$$

where the Kronecker delta $\delta_{ij} = 1$ for $i = j$ and 0 otherwise, assuming for simplicity no demographic covariances among the vital rates. We approximate n_i using the proportion in the stable age distribution in the average environment, $n_i \cong u_i N$. Substituting these covariances into equation 3.9b then gives the infinitesimal mean and variance of the diffusion approximation for the log of total population size, $X = \ln N$,

$$M(X) = \ln \lambda - \frac{V(X)}{2} \quad \text{and} \quad V(X) = \sigma_e^2 + \frac{\sigma_d^2}{N} \qquad 3.10a$$

where the environmental and demographic variances are defined as

$$\sigma_e^2 = \sum_i \sum_j \left(\frac{\partial \ln \lambda}{\partial \pi_i} \right) \left(\frac{\partial \ln \lambda}{\partial \pi_j} \right) \text{Cov}[\mu_i, \mu_j] \quad \text{and} \quad \sigma_d^2 = \sum_i \left(\frac{\partial \ln \lambda}{\partial \pi_i} \right)^2 \frac{\text{Var}[\delta_i]}{u_i}$$
$$3.10b$$

and the initial condition is again $X(0) = \ln(\mathbf{vn}(0)/(\mathbf{vu}))$.

Equations 3.10a are analogous to formulas 1.3 and 1.4 for a population with no age structure. Simulations show that this univariate diffusion process with three parameters λ, σ_e^2, and σ_d^2, accurately approximates the distribution of time to extinction in density-independent age-structured populations with both demographic and environmental stochasticity (Engen *et al.*, 2003*b*). This helps to justify the analysis of extinction and quasi-extinction using simplified models with no age structure (Chapters 2 and 5). We now investigate the influence of age structure on the temporal pattern of population fluctuations around a density-dependent equilibrium.

3.3 Time series analysis of population fluctuations

Populations of terrestrial vertebrates often fluctuate around an equilibrium or average size with a small or moderate coefficient of variation (see Fig. 1.1 and Table 1.1). For such populations the theory of stationary time series can be applied to analyze the pattern of temporal fluctuations (Kendall *et al.*, 1983; Box *et al.*, 1994; Chatfield, 1996). A stationary time series can be characterized by its autocorrelation function, known as the correlogram, and by its spectral density function, known as the power spectrum, which is the Fourier transform of the autocorrelation function. The power spectrum describes the proportion of the total variance in population size attributable to cycles of random (uniformly distributed) phase as a function of their frequency (Chatfield, 1996). For most populations the power spectrum displays a substantial 'red shift' toward low frequencies in comparison with the flat power spectrum associated with 'white noise' that describes populations with no temporal autocorrelation (Pimm, 1991; Ariño and Pimm, 1995).

Time series from eight vertebrate populations (six birds and two mammals) were chosen based on having three or more decades of accurate annual census data (Fig. 3.4). Years when a complete census was not undertaken were excluded from the analysis of the two mammals, the Chamois and Soay Sheep, and two of the birds, the Mute Swan and South Polar Skua. The population of Soay Sheep on Hirta Island is closed to immigration, and counts include only adult females (≥ 1 year old), which far outnumber adult males, with an average sex ratio of 4.5 adult females per adult male. The Chamois population in the Swiss National Park in southwestern Switzerland is nearly closed, and the counts have much higher accuracy than usual for ungulates, but include calves and juveniles. Counts of the Great Tit and Blue Tit at Ghent, Belgium, and the Tufted Duck at Engure Marsh, Latvia, include the total adult population (≥ 1 year old). High proportions of the tit populations reproduce in nest boxes, facilitating accurate counts, but there is considerable exchange of individuals with other populations. The Grey Heron counts are for breeding adults (≥ 2 years old) in southern Britain, which compose a relatively closed population. Counts of the Mute Swan on the Thames, England, are for the total population minus fledglings. Some of the Mute Swan annual counts may be biased (Birkhead and Perrins, 1986) and fledglings were wing-clipped during the counts to reduce emigration (Cramp, 1972); this series is included mainly for illustrative purposes because of its length. The South Polar Skua population at Pointe Géologie archipelago, Terre Adélie, Antarctica, has a significant input of recruits from outside the archipelago. The strong territorial behavior of adults helps to ensure that all birds in the archipelago are metal and color-ringed and the counts of breeding adults are exact.

Formulas for statistical analysis of time series typically assume that the length of the series, L, is much greater than the time lags of interest, τ (Chatfield, 1996). However, this often does not apply to ecological time series that usually have lengths measured in decades, because a density-dependent life history can produce population autocorrelations on time scales much longer than a generation (see after eqs 1.9). In tests of statistical significance for estimated autocorrelations we therefore employ the sampling variance $1/(L - \tau)$ (Kendall *et al.*, 1983), instead of the commonly used approximation $1/L$ (Chatfield, 1996). Thus in Figure 3.4 the width of the confidence intervals on the correlograms increases with the lag τ. Consistent statistical estimation of the power spectrum from time series data requires that in the Fourier transform of the empirical correlogram short time lags should be weighted much more heavily than long time lags, using a 'lag window' (Box *et al.*, 1994; Chatfield, 1996).

The 20-year cycle in Grey Heron abundance suggested by the correlogram may be largely an artifact of a single catastrophic winter in 1963 that caused the Grey Heron population to fall by nearly half (Fig. 3.4). For the Mute Swan, significant negative autocorrelations at long time lags might be caused by a linear trend (Chatfield, 1996) or a long-term cycle in the data. However, random long-term fluctuations that appear

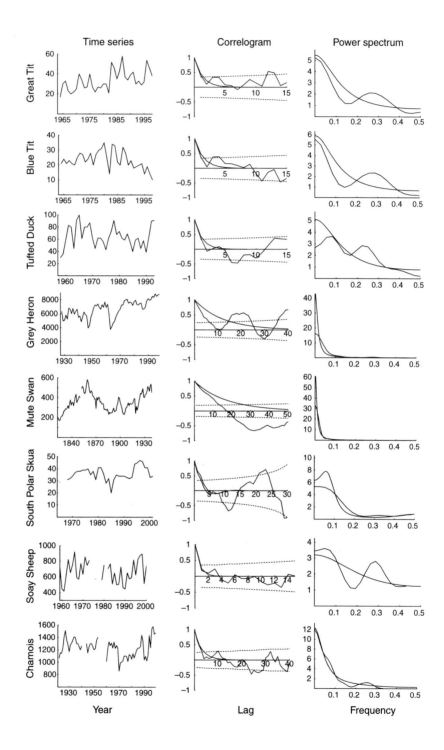

| Time series | Correlogram | Power spectrum |

cyclic in time series of limited duration are typical of species with weak density dependence (see Fig. 3.5 top sample path).

Populations of each of the eight species in Figure 3.4 have power spectra that display the usual red shift toward low frequencies, as discussed by Pimm (1991) and Ariño and Pimm (1995). These authors also emphasize that the observed (or empirical) variance in census population size tends to increase with the period of observation, over time scales as long as decades. This occurs because of either autocorrelation or nonstationarity, both of which cause extreme population sizes to be undersampled in short time series. Nonstationarity in the form of a sustained population trend produces large negative autocorrelations at long time lags (Chatfield, 1996). For a stationary time series of population sizes, $N(t)$, with expectation μ_N and variance σ_N^2, censused completely during L consecutive years, the expected value of the empirical variance, $s_N^2 = (1/L) \sum_{t=1}^{L} [N(t) - \bar{N}]^2$ where $\bar{N} = (1/L) \sum_{t=1}^{L} N(t)$, can be derived by adding $\mu_N^2 - \mu_N^2 = 0$ to the right side, and rearranging,

$$
\begin{aligned}
E[s_N^2] &= E\left\{ \frac{1}{L} \sum_{t=1}^{L} [N(t)]^2 - \mu_N^2 \right\} - \left\{ E\left[\frac{1}{L} \sum_{t=1}^{L} N(t) \right] - \mu_N \right\}^2 \\
&= \frac{1}{L} \sum_{t=1}^{L} \sigma_N^2 - \frac{1}{L^2} \sum_{t=1}^{L} \sum_{u=1}^{L} \rho(t-u) \sigma_N^2 \\
&= \left[1 - \frac{1}{L} - \frac{2}{L} \sum_{\tau=1}^{L} \left(1 - \frac{\tau}{L} \right) \rho(\tau) \right] \sigma_N^2
\end{aligned}
\tag{3.11}
$$

where $\rho(\tau)$ is the population autocorrelation for lag τ, such that $E[N(t)N(t-\tau)] - \mu_N^2 = \rho(\tau)\sigma_N^2$ with $\rho(\tau) = \rho(-\tau)$ and $\rho(0) = 1$. Thus for a stationary process the expected empirical variance in a time series of length L depends directly on the autocorrelation function. In the absence of autocorrelation, with L independent observations of population size, equation 3.11 reduces to the standard formula, $(1 - 1/L)\sigma_N^2$, for the expected empirical variance in a random sample of L observations from a hypothetical infinite universe of observations with variance σ_N^2 (Stuart and Ord, 1994). In general, $E[s_N^2]$ closely approaches its asymptotic value σ_N^2 only when the length of the time series greatly exceeds the net autocorrelation, $L \gg 1 + 2 \sum_{\tau=1}^{L} \rho(\tau)$. A slow increase in empirical variance with increasing length

---

Fig. 3.4   Time series, empirical correlogram, and power spectrum for annual census of adult population in six avian species and two large mammalian species. Theoretical autocorrelation function and power spectrum (smooth curves) were calculated for each species from estimated autoregression coefficients in the stage-structured life history model (eq. 3.15, Table 3.1) using analytical methods in Box *et al.* (1994). (Modified from Lande *et al.*, 2002.)

of time series is a consequence of positive autocorrelations, which can be caused by weak density dependence and/or long generation time, the same factors responsible for red shift of the power spectrum (see after eqs 1.9, Figs 3.4, 3.5 and Tables 3.1, 3.2). The difference between σ_N^2 and $E[s_N^2]$ equals $Var[\bar{N}]$, the sampling variance in the empirical mean of a time series, which can be substantial for series with high positive autocorrelations that fail to satisfy the foregoing inequality.

3.4 Estimation of density dependence from population time series

Ecological studies of density dependence began with debate about its occurrence (Andrewartha and Birch, 1954; Tamarin, 1978), and progressed to detection and measurement in experimental and observational studies (Hassell and May, 1990; Harrison and Cappuccino, 1995), including statistical inference from population time series (Bulmer, 1975; Pollard et al., 1987; Hanski et al., 1993; Dennis and Taper, 1994; reviewed by Turchin, 1995). Detection and estimation of density dependence is complicated because it usually operates with a time delay due to intrinsic factors in individual development and life history (May, 1973, 1981; MacDonald, 1978; Renshaw, 1991; Nisbet, 1997; Jensen, 1999; Claessen et al., 2000), and extrinsic factors in an autocorrelated environment (Williams and Liebhold, 1995; Berryman and Turchin, 1997), including ecological interactions among species (Turchin, 1990, 1995; Royama, 1992; Turchin and Taylor, 1992; Kaitala et al., 1997; Ripa et al., 1998; Hansen et al., 1999). The life history of a species may largely determine the relative importance of intrinsic and extrinsic factors in contributing to time delays in population dynamics. For short-lived species with high population growth rates, such as some insects, ecological interactions may best explain time delays longer than a generation (Turchin, 1990, 1995; Royama, 1992). For long-lived species with low population growth rates, such as large vertebrates, most time delays may occur within a generation because of the development time for individuals to reach reproductive ages (Jensen, 1999; Coulson et al., 2001a; Thompson and Ollason, 2001).

Although it is well known that life history can create time delays in population dynamics (e.g. eq. 3.1c, references above), this basic fact was not incorporated in general methods for detecting and estimating density dependence from population time series (Bulmer, 1975; Pollard et al., 1987; Turchin, 1990, 1995; Royama, 1992; Turchin and Taylor, 1992; Hanski et al., 1993; Dennis and Taper, 1994; Zeng et al., 1998) until quite recently. Following Lande et al. (2002), we analyze populations with density-dependent stage-structured life histories that reproduce at discrete annual intervals, with small or moderate population fluctuations around a stable equilibrium. As previously discussed in Chapter 1, vertebrate species with mean adult body mass greater than 1 kg usually have $r_{max} \leq 0.1$ per year (Charnov, 1993), and even highly fecund species, such as many fish, insects, and plants, have maximum population growth rates that are limited by high density-independent mortality (Myers et al., 1999). Such populations typically show damped fluctuations around a stable equilibrium (May, 1981) with a small or moderate coefficient of variation (Table 1.1; Pimm and Redfearn, 1988; Pimm, 1991).

Employing a life history with age at maturity α, and stochasticity and density dependence in adult recruitment and mortality rates, we derive a linearized autoregressive equation with time delays from 1 to α years. Contrary to current interpretations (e.g. Turchin, 1990, 1995; Royama, 1992; Turchin and Taylor, 1992; Zeng *et al.*, 1998), the coefficients corresponding to different time delays in the autoregressive dynamics do not simply measure delayed density dependence, but also depend on the life history. The theory indicates that the total density dependence in the life history, D, can be estimated from the sum of the autoregression coefficients. We apply this theory to time series for terrestrial vertebrate populations depicted in Figure 3.4.

3.4.1 Quantitative definition of density dependence

A major impediment to understanding density dependence has been the lack of a general quantitative definition that would allow comparisons among species with different life histories and forms of density dependence (Murdoch, 1994). We now derive such a definition. First consider a simple deterministic model of a population with no age structure, in which individuals mature at age 1 year, reproduce, and then die, as for univoltine insects or annual plants with no seed bank. The same dynamics occurs for a population in which individuals mature at age 1 year and adults have age-independent annual survival and reproduction. Denoting the population size in year t as $N(t)$, the dynamics can be written as $N(t) = \lambda[N(t-1)]N(t-1)$ where $\lambda[N(t-1)]$ is the density-dependent multiplicative growth rate (see Chapter 1). Let the equilibrium or average population size be K and denote the deviation from equilibrium as $x(t) = N(t) - K$. Taylor expansion of λ gives the linearized dynamics $x(t) = (1-\gamma)x(t-1)$ where the rate of return to equilibrium, $\gamma = -(\partial \ln \lambda / \partial \ln N)_K$, is the negative elasticity of population growth rate with respect to change in population density at equilibrium (de Kroon *et al.*, 1986; Caswell, 2001, p. 226).

This definition of density dependence can be generalized to age-structured populations with density-dependent stochastic dynamics as follows. First, note that the linearized dynamics in the simple model without age structure also applies asymptotically to a deterministic age-structured population where λ is the asymptotic multiplicative growth rate of the population per year. A general age-structured life history model with density dependence in age-specific fecundity and first year survival (Lande *et al.*, 2002) indicates that the total density dependence, D, should be defined as the negative elasticity of population growth rate per generation, λ^T, with respect to change in population density of adults, evaluated at equilibrium. The generation time, T, is defined in equation 3.4. Using $\ln \lambda^T = T \ln \lambda$ and that at equilibrium $\lambda = 1$ or $\ln \lambda = 0$, we find

$$D = -\left(\frac{\partial \ln \lambda^T}{\partial \ln N}\right)_K = -\left(T\frac{\partial \ln \lambda}{\partial \ln N} + \ln \lambda \frac{\partial T}{\partial \ln N}\right)_K = -\left(T\frac{\partial \ln \lambda}{\partial \ln N}\right)_K. \qquad 3.12$$

Thus with age structure the asymptotic rate of return to equilibrium is the total density dependence in the life history divided by the generation time at equilibrium, $\gamma = D/T$. Remarkably, in the stage-structured life history analyzed below, this definition of total density dependence (eq. 3.12) still applies despite the occurrence of density dependence in juvenile and adult survival as well as recruitment.

3.4.2 Density-dependence in a stage-structured life history

Many wild bird and mammal populations have life histories in which adult annual survival and reproductive rates are roughly constant and independent of age (Deevey, 1947; Gaillard *et al.*, 1994; Nichols *et al.*, 1997; Loison *et al.*, 1999). In such populations the great majority of individuals die before reaching the age of senescence, which can therefore be neglected. We assume that all density dependence is exerted by the adult population density, which applies, at least approximately, if juveniles do not compete with adults or if adults are long-lived and juveniles compose a small fraction of the population. Denoting the age at first breeding as α and the number of adults (of age $\geq \alpha$) in year t as $N(t)$, the stochastic density-dependent dynamics are described by the nonlinear recursion

$$N(t) = s(N, t-1)N(t-1) + f(N, t-\alpha) \prod_{i=1}^{\alpha-1} s_{\alpha-i}(N, t-i)N(t-\alpha). \quad 3.13$$

The probability of adult annual survival is s. The adult annual recruitment rate is the product of annual fecundity (female offspring per adult female per year) times first year survival, f, and the probabilities of annual survival from age i to $i+1$ during the juvenile stages, s_i. These vital rates are density and time dependent and population sizes in these functions have the same time dependence as the functions themselves. Environmental and demographic stochasticity affect the vital rates through additive perturbations without changing their expected values at a given adult population size, $\bar{s}(N)$, $\bar{f}(N)$ and $\bar{s}_\tau(N)$. The total density dependence in this life history can be derived by implicit differentiation of the deterministic Euler–Lotka equation 3.8. Finding $\partial\lambda/\partial N$, evaluating the result at equilibrium, with $N = K$ and $\lambda = 1$, and finally using the generation time for this stage-structured life history at equilibrium in the average environment, $T = \alpha + \bar{s}/(1 - \bar{s})$ (eq. 3.8), gives

$$D = -\left(\frac{\partial \ln \bar{f}}{\partial \ln N} + \sum_{\tau=1}^{\alpha-1} \frac{\partial \ln \bar{s}_\tau}{\partial \ln N} + \frac{\bar{s}}{1 - \bar{s}} \frac{\partial \ln \bar{s}}{\partial \ln N}\right)_K = -\left(\frac{\partial \ln(\bar{\phi}/\bar{\mu})}{\partial \ln N}\right)_K \quad 3.14$$

where $\bar{\phi}(N) = \bar{f}(N) \prod_{i=1}^{\alpha-1} \bar{s}_i(N)$ is the adult recruitment rate in the average environment, and $\bar{\mu} = 1 - \bar{s}$ is the adult mortality rate. Thus density dependence in the stage-structured life history can be measured by the negative elasticity of the ratio of adult recruitment rate to adult mortality rate with respect to changes in adult population density at equilibrium. Denoting equilibrium values as $\hat{s} \equiv \bar{s}(K)$ and $\hat{\phi} \equiv \bar{\phi}(K)$, the Euler–Lotka equation 3.8 with $N = K$ and $\lambda = 1$ reveals that at equilibrium the adult recruitment rate equals the adult mortality rate, $\hat{\phi} = 1 - \hat{s} = \hat{\mu}$.

Expanding the time-dependent vital rates in equation 3.13 in Taylor series around the deterministic equilibrium adult population size, K, with deviations from equilibrium denoted as $x(t) = N(t) - K$, yields the linearized autoregression for small fluctuations, involving time delays from 1 to α years,

$$x(t) = \sum_{\tau=1}^{\alpha} b_\tau x(t - \tau) + \xi(t) \qquad\qquad 3.15$$

with constant coefficients

$$b_1 = \left[1 + \left(\frac{\partial \ln \bar{s}}{\partial \ln N} \right)_K \right] \hat{s} + \left(\frac{\partial \ln \bar{s}_{\alpha-1}}{\partial \ln N} \right)_K \hat{\phi}$$

$$b_\tau = \left(\frac{\partial \ln \bar{s}_{\alpha-\tau}}{\partial \ln N} \right)_K \hat{\phi} \quad \text{for } \tau = 2, \ldots, \alpha - 1 \quad \text{and}$$

$$b_\alpha = \left[1 + \left(\frac{\partial \ln \bar{f}}{\partial \ln N} \right)_K \right] \hat{\phi}.$$

The noise term $\xi(t)$ has mean zero, but depends on temporal fluctuations in vital rates operating over time delays up to α years (Lande *et al.*, 2002). Hence for species with $\alpha > 1$, even in the absence of environmental autocorrelation in the vital rates, the noise $\xi(t)$ will be autocorrelated if the vital rates operating at different time delays are cross-correlated at a given time. Autocorrelation of physical and biotic environments also have been discussed as causes of autocorrelated noise in autoregression analysis of delayed density dependence (Williams and Liebhold, 1995; Berryman and Turchin, 1997).

Autocorrelated noise presents a potential complication for the measurement of density dependence from population time series. The following autoregression analyses of population time series, like previous studies (e.g. Turchin, 1990, 1995; Royama, 1992; Turchin and Taylor, 1992; Zeng *et al.*, 1998), assume that autocorrelation of the noise is negligible. Residuals from the autoregressions showed no significant autocorrelations, suggesting not only a negligible environmental autocorrelation (see Chapter 1, Table 1.4), but also that environmental covariance of vital rates operating at different time lags is small. Long-term life history studies of vertebrate species often show that reproduction and early juvenile survival are much more variable among years than adult mortality (Gaillard *et al.*, 1998, 2000; Sæther and Bakke, 2000), as observed in the Tufted Duck (Blums *et al.*, 1996), Grey Heron (North and Morgan, 1979), and Mute Swan (Cramp, 1972; Bacon and Perrins, 1991), which would limit the scope for environmental covariance of vital rates in the stage-structured model.

The theoretical autocorrelation function and power spectrum for the stage-structured life history were derived using analytical methods for linear time series in Box *et al.* (1994, Lande *et al.*, 2002). Figure 3.5 illustrates for a life history with

$\alpha = 4$ how the strength of density dependence (in fecundity only) influences the sample paths, and the theoretical autocorrelation function and power spectrum, also demonstrating how the empirical variance in population size is expected to increase with length of the time series.

Statistical analysis of population dynamics commonly is carried out using $\ln N$ rather than N (Royama, 1992; Turchin, 1995). The form of the linearized autoregression (the vector of autoregression coefficients) is exactly the same for the dynamics of $\ln N$ as it is for N. This can be shown by dividing both sides of equation 3.15 by K and noting that for small fluctuations $x/K \cong \ln(1 + x/K) = \ln(N/K) = \ln N - \ln K$.

3.4.3 Estimating density dependence

The autoregression coefficients can be estimated from the population autocorrelation function for an observed population time series. Time series at least an order of magnitude longer than the number of autoregression coefficients are usually required for accurate estimation. Neglecting end effects due to finite length of the series, the maximum likelihood estimator of the autoregression coefficients is identical to the least squares estimator for a standard regression, given by inverting the Yule–Walker equations, $\mathbf{b} = \mathbf{P}^{-1}\boldsymbol{\rho}$ where $\mathbf{b}$ and $\boldsymbol{\rho}$ are column vectors with elements b_τ and $\rho(\tau)$ for $\tau = 1, \ldots, \alpha$ and $\mathbf{P}$ is the population autocorrelation matrix with elements $P_{ij} = \rho(|i - j|)$ for $i, j = 1, \ldots, \alpha$ with $\rho(0) = 1$ (Kendall et al., 1983). For example, for $\alpha = 1$ the sole autoregression coefficient is estimated by $b_1 = \rho(1)$. These estimators of the autoregression coefficients are biased because population sizes at a given time enter the autoregression as both dependent and independent variables, in contrast to standard regression, which assumes independent observations (Bulmer, 1975; Dennis and Taper, 1994; Caswell, 2001, p. 142). This time series bias can be estimated and removed using computer simulations. Starting with the initial estimates of the autoregression coefficients, we estimated the bias by performing 10,000 stochastic simulations of the time series, re-estimating the autocorrelations and the b_τ in each simulation. The mean values of the autoregression coefficients b_τ from these simulations were then used in another set of 10,000 simulations to make a second bias correction. Each set of simulations was performed with the same random number seed at the beginning of the set, giving repeatable sequences of the stochastic variables. With this procedure, convergence of the bias corrected values usually is rapid, within a few iterations. Standard errors for the b_τ were estimated by parametric bootstrapping, simulating the processes using the bias corrected estimates. Computer simulations also can be used for significance testing and to obtain confidence intervals (Lande et al., 2002). Residuals from the autoregression revealed no significant autocorrelations, justifying the approximation of independent errors in estimation and significance testing.

In general there are α autoregression coefficients, and summing these using equations 3.14 and 3.15 with $\hat{s} + \hat{\phi} = 1$ reveals that we can estimate

$$\hat{\mu}D = 1 - \sum_{\tau=1}^{\alpha} b_\tau. \qquad 3.16$$

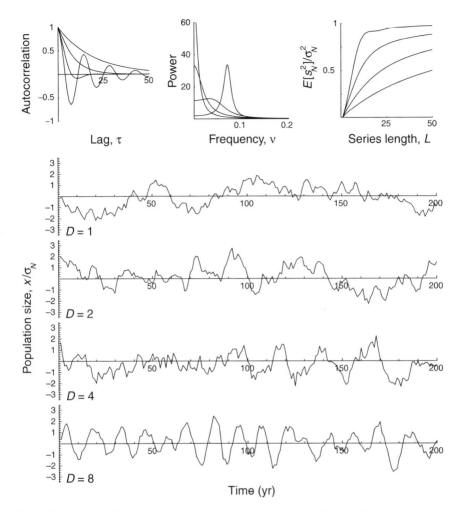

Fig. 3.5 Autocorrelation functions, $\rho(\tau)$, power spectra, and simulated sample paths for small fluctuations in adult population size around a density-dependent equilibrium in the stage-structured life history model. All density regulation is exerted by the number of adults on the adult fecundity times first year survival, f, and all stochasticity occurs in this vital rate alone. The expected empirical variance of population size, as a proportion of the stationary variance, is shown as a function of the length of the sample series. Power spectra, as a function of frequency in cycles per year, are normalized to unit area. Population sizes are plotted as deviations from the mean, standardized to unit variance. Age at maturity is $\alpha = 4$ years. Adult vital rates are independent of age, with adult annual mortality rate $\mu = 0.05$. Different lines in each panel correspond to values of density dependence in fecundity $D = -(\partial \ln \bar{f}/\partial \ln N)_K = 1, 2, 4,$ or 8. Weak density dependence produces positive autocorrelation over long time lags and a red shift in the power spectrum. Sufficiently strong density dependence creates negative autocorrelations, and a tendency to cycle in the sample paths, corresponding to an internal peak in the power spectrum at a frequency that increases with the strength of density dependence.

This theory for estimating density dependence was applied to the population time series in Figure 3.4 with results displayed in Tables 3.1 and 3.2.

Contrary to prevalent interpretations in the literature the autoregression coefficients in equation 3.15 do not directly reveal the strength of delayed density dependence in population dynamics. The autoregression coefficient at a given time delay depends both on density dependence in the vital rates and on the life history parameters. For example, in a species with age at maturity $\alpha > 1$, with no density dependence in $\bar{f}$ we see that $b_\alpha = \hat{\phi}$, or alternatively, with no density dependence in adult or subadult annual survival rates, s and $s_{\alpha-1}$ we see that $b_1 = \hat{s}$. It is also instructive to consider a species with $\alpha = 1$, having only a single autoregression coefficient $b_1 = 1 - \hat{\mu}D$ from which we can estimate $\hat{\mu}D = 1 - b_1$; if $b_1 = 0$ then the population autocorrelations for all time lags are zero (except $\rho(0) = 1$) corresponding to a flat power spectrum or white noise process for the population. This would entail strong density dependence, $D = 1/\hat{\mu}$, despite the regression explaining none of the total variance, $R^2 = 0$. The Soay Sheep population approaches this situation, having a rather low R^2 (Table 3.1). Thus statistical significance of autoregression coefficients is not a valid criterion for the detection of density dependence.

For the Grey Heron (North and Morgan, 1979), Chamois (Loison et al., 1994), and Mute Swan (Bacon and Perrins, 1991), with $\alpha > 1$, the estimated adult survivorships from life history studies are, respectively, $\hat{s} = 0.70, 0.96$, and 0.77. These are not significantly different from the respective autoregression estimates $b_1 = 0.93, 0.73$, and 0.85, and subsequent regression coefficients up to $b_{\alpha-1}$ do not differ significantly from zero. Thus for these species we can not reject the hypothesis of no density dependence in survival. Even assuming this to be true, discrepancies between the life history and autoregression estimates of $\hat{s}$ could still occur due to different years studied.

For the South Polar Skua the average age of first reproduction is 6 years (Jouventin and Guillotin, 1979; H. Weimerskirch, unpubl. data). Estimates of the first three autoregression coefficients and the total density dependence in the life history are rather insensitive to values of α ranging from 3 to 5 years, but for $\alpha = 6$ the estimated autoregression coefficients became unreliable, with large standard errors, apparently because there were too many coefficients relative to the length of the series for accurate estimation (Lande et al., 2002). We used the results for $\alpha = 5$ in Tables 3.1 and 3.2. The estimated adult survivorship from life history studies (H. Weimerskirch, unpublished data) $\hat{s} = 0.85$ is significantly larger than the first autoregression coefficient $b_1 = 0.59$ and the intermediate autoregression coefficients between b_1 and b_α include some substantial (although not significant) estimates, suggesting appreciable density dependence in survivorship beyond the first year in this species, possibly because of the strong territoriality.

Significant total density dependence was found in six of the eight populations (Table 3.2). This is not surprising because some form of population density regulation is required to maintain a low coefficient of variation over a long time in a fluctuating environment. Comparing the strength of total density dependence, D, between species requires correcting the estimates of $\hat{\mu}D$ by dividing by the adult annual mortality rate

Table 3.1 Bias-corrected autoregression coefficients, b_τ ($\pm$SE), and proportion of variance explained by the autoregression, R^2, for population time series fitted to the stage-structured life history model (eq. 3.15), with age of first reproduction α obtained from the literature (Owen, 1960; Bacon and Andersen-Harild, 1989; Dhondt et al., 1990; Loison et al., 1994; Blums et al., 1996; Clobert et al., 1998; Coulson et al., 2001a; H. Weimerskirch unpubl. data)

Species	Years	CV	α	b_1	b_2	b_3	b_4	b_5	R^2	Reference
Great Tit	35	0.30	1	0.465 ($\pm$0.159)					0.16	E. Matthysen (unpubl.)
Blue Tit	35	0.26	1	0.493 ($\pm$0.156)					0.15	E. Matthysen (unpubl.)
Tufted Duck	36	0.27	1	0.436 ($\pm$0.156)					0.14	Blums et al. (1993)
Soay Sheep	37	0.21	1	0.230 ($\pm$0.161)					0.03	T. Clutton-Brock (unpubl.)
Grey Heron	71	0.18	2	0.926 ($\pm$0.127)	−0.006 ($\pm$0.119)				0.73	British Trust for Ornithology
Chamois	69	0.12	3	0.728 ($\pm$0.133)	−0.057 ($\pm$0.157)	0.045 ($\pm$0.130)			0.46	F. Filli (unpubl.)
Mute Swan	116	0.26	4	0.850 ($\pm$0.100)	0.078 ($\pm$0.127)	−0.002 ($\pm$0.125)	0.011 ($\pm$0.095)		0.80	Cramp (1972)
S. Polar Skua	35	0.16	5	0.587 ($\pm$0.207)	0.339 ($\pm$0.221)	−0.259 ($\pm$0.211)	−0.078 ($\pm$0.213)	0.014 ($\pm$0.188)	0.29	H. Weimerskirch (unpubl.)

Table 3.2 Bias-corrected estimates of adult annual mortality rate times total density dependence, $\hat{\mu}D$ (95% confidence interval), from eq. 3.16. Total density dependence is estimated from $D = [1 - \Sigma_i b_i]/(1 - s)$ where s is the adult annual survival rate obtained from the life history studies cited. Density dependence per year is calculated from $\gamma = D/T$ where the generation time T (eq. 3.8) is obtained from α (Table 3.1) and s

Species	$\hat{\mu}D$	s	D	T	γ	Reference for s
Great Tit	0.535** (0.184, 0.879)	0.46	0.99	1.85	0.54	Clobert et al. (1988)
Blue Tit	0.507** (0.157, 0.849)	0.49	0.99	1.96	0.51	Dhondt et al. (1990)
Tufted Duck	0.564** (0.221, 0.907)	0.65	1.61	2.86	0.56	Blums et al. (1996)
Soay Sheep	0.770*** (0.415, 1.107)	0.84	4.81	6.25	0.77	T. Coulson and T. Clutton-Brock (unpubl.)
Grey Heron	0.081 (0.000, 0.246)	0.70	0.27	4.33	0.06	North and Morgan (1979)
Chamois	0.283** (0.080, 0.573)	0.96	7.08	27	0.26	Loison et al. (1994)
Mute Swan	0.062 (0.000, 0.184)	0.85	0.41	9.67	0.04	Bacon and Perrins (1971)
S. Polar Skua	0.397*** (0.103, 1.114)	0.85	2.65	10.67	0.25	Jouventin and Guillotin (1979), H. Weimerskirch (unpubl.)

** $P < 0.01$, *** $P < 0.001$ for hypotheses that $\hat{\mu}D > 0$ by one-tailed test.

obtained from life history studies, as shown in Table 3.2. The total density dependence appears to be strong in all of the populations in which significant estimates were obtained, especially for the two mammals. Strong density dependence of recruitment (fecundity and/or juvenile survival) is often observed in ungulates (Grenfell *et al.*, 1992; Sæther, 1997; Gaillard *et al.*, 2000). Lack of significant density dependence for the Grey Heron and especially for the Mute Swan, which had the longest time series, suggests that density dependence is weak in these species.

In the six avian species $\bar{D} = 1.16$ so that on average near equilibrium a given proportional increase in adult population density, N, produces roughly the same proportional decrease in multiplicative growth rate of the population per generation, λ^T, whereas in the two large mammalian species with $\bar{D} = 5.95$ the average effect is several times larger. Despite evidence of strong density dependence in components of the life history in some of the species, especially in recruitment of long-lived species, in each case the generation time exceeds the total density dependence in the life history, so that the expected annual rate of return to equilibrium is always less than unity, $\gamma = D/T < 1$ (Table 3.2), or in other words, near equilibrium a proportional increase in N produces a smaller proportional decrease in λ. The expected dynamics of adult population size are therefore undercompensated (Begon *et al.*, 1996, pp. 239–40). However, in contrast to models without age structure, undercompensation is not a sufficient condition to describe the stability properties of an age- or stage-structured life history with intrinsic time delays. A general condition for the expected dynamics to be smoothly damped is that the life history model fitted to time series data should not display negative autocorrelations nor an internal peak in the power spectrum. Inspection of the model autocorrelation functions and power spectra in Figure 3.4 show that all but one of the bird and mammal populations satisfy this condition, implying that their expected dynamics are smoothly damped, rather than oscillatory, cyclic, or chaotic. The sole exception is the South Polar Skua, which has the longest age at maturity (Table 3.1) and is therefore most susceptible to the destabilizing time delays in the stage-structured life history model; alternatively, the apparent internal cycle in the model may also be an artifact of externally driven long-term fluctuations in ocean conditions.

This method for estimating density dependence can be extended to populations with large fluctuations around a (stable or unstable) equilibrium by fitting a particular form of the general nonlinear stage-structured life history model (eq. 3.13) to a population time series. The fitted model can then be used to evaluate analytically or numerically the total density dependence in the life history, $D = -\partial \ln \lambda^T / \partial \ln N$, at the equilibrium point or averaged over the (quasi-)stationary distribution of N.

3.5 Summary

Classical deterministic results for density-independent growth of an age-structured population can be extended to include demographic and environmental stochasticity.

In a constant environment, a population asymptotically approaches a stable age distribution with a constant multiplicative growth rate, λ, but the total reproductive value always increases at a constant rate. In a random environment, the expected population size increases at the asymptotic growth rate in the average environment, but the long-run growth rate is reduced by environmental stochasticity. Demographic and environmental variances in an age-structured population are defined in terms of components of variances and covariances of age-specific vital rates and the sensitivities of λ to small changes in the vital rates.

The total density dependence in a life history is defined as the negative elasticity of population growth rate per generation with respect to population density. We analyze a stage-structured life history model with age at maturity α, and adult annual mortality and reproductive rates independent of age, with all density dependence exerted by the adult age class on all the vital rates. For small or moderate fluctuations in adult population size around an equilibrium or average size, we derive a linear autoregressive equation with time delays from 1 to α years. This demonstrates that a nonzero autoregression coefficient at a given time delay does not generally measure delayed density dependence, but also can be produced by intrinsic delays in the life history. The total density dependence in the life history can be estimated using the sum of the autoregression coefficients. Applying this theory to time series of terrestrial vertebrate populations reveals that, despite evidence for strong density dependence in some components of the life history, the expected dynamics of adult population size are undercompensated, $\gamma < 1$. Thus, near equilibrium a proportional increase in adult population density is expected to produce a smaller proportional decrease in λ. Inspection of model autocorrelation functions and power spectra calculated by fitting population time series to the stage-structured life history show that the expected dynamics usually are smoothly damped rather than cyclic or chaotic. Long autocorrelation and a red-shifted power spectrum are produced by a combination of weak density dependence and long generation times.

4

Spatial structure

4.1 Classical metapopulations

Population subdivision and spatial heterogeneity of the environment can exert a strong influence on population dynamics and extinction risk. Spatial subdivision of a population arises from barriers to dispersal, habitat fragmentation, social grouping, or simply from geographic distance greater than the scale of individual dispersal. Local populations, also called subpopulations or demes, undergo independent demographic stochasticity and may experience different environments. Demographic and environmental stochasticity within local populations creates a risk of local extinction, and dispersal to vacant habitat can lead to successful colonization. Levins (1969, 1970) introduced the concept of a metapopulation composed of a set of geographically distinct local populations occupying patches of suitable habitat. The subpopulation on each habitat patch is subject to local extinction but may be recolonized by individual dispersal from other occupied patches. Levins developed his model to describe the control and eradication of pest species, but its main application now is in conservation biology. Levins' model and subsequent extensions (Hanski and Gilpin, 1997; Hanski, 1999) have proven important in understanding the dynamics of natural populations as habitat destruction and fragmentation has reduced many landscapes to islands of suitable habitat, surrounded by a sea of unsuitable habitat. Related concepts have helped to focus attention on the consequences of habitat heterogeneity for population dynamics. The critical patch size describes the minimum area of suitable habitat that can sustain an isolated population in the face of random individual dispersal into surrounding unsuitable regions (Skellam, 1951; Kierstead and Slobodkin, 1953; Okubo, 1980). Population sources and sinks, respectively, describe areas of net emigration and immigration (Pulliam, 1988; Pulliam and Danielson, 1992).

Levins (1969, 1970) proposed that regional persistence of a population in a fragmented habitat may occur through a balance between random local extinction and colonization. For simplicity, he assumed that each patch of suitable habitat is either occupied or unoccupied, with constant rates of local extinction, e, and colonization, c, per occupied patch. Although local extinction and colonization of each patch of suitable habitat is a stochastic process, in a very large, effectively infinite, region the rate of change in the proportion of suitable habitat patches that are occupied becomes an essentially deterministic process. This and other classical metapopulation models are spatially implicit, in that the geographic location of the habitat patches is ignored. The proportion of suitable habitat patches occupied by a species, p, obeys the continuous-time differential equation

$$\frac{dp}{dt} = cp(1 - p) - ep \qquad 4.1$$

The colonization term is proportional to both p and $1 - p$ because colonists must originate from an occupied patch and colonize an unoccupied patch of suitable habitat. At equilibrium $dp/dt = 0$ and the occupancy of suitable patches is $p = 1 - e/c$. In this model the metapopulation persists, with $p > 0$ at equilibrium, only if the local extinction rate is less than the colonization rate, $e < c$. Although this conclusion is intuitively obvious, it has the important implication that a metapopulation can become extinct deterministically even in the presence of suitable habitat. The balance between local extinction and colonization required for metapopulation persistence means that preserving currently unoccupied suitable (or temporarily unsuitable) habitat is potentially as important as currently occupied habitat to prevent regional extinction. This concept has only recently begun to reach the awareness of many environmental managers. The question of how much suitable habitat is necessary for metapopulation persistence is addressed in the following extension of Levins' model to territorial species.

Levins' model and most extensions of it make several simplifying assumptions. The most critical assumptions involve ignoring the local population dynamics, so that each patch of suitable habitat is either occupied or unoccupied, with rates of local extinction and colonization per extant subpopulation assumed to be constant. However, these assumptions are actually inconsistent with basic population dynamics as individual dispersal and the stochastic dynamics of local populations actually determine the rates of local extinction and colonization. This chapter reviews models of spatially subdivided populations with explicit stochastic dynamics of local populations. We consider metapopulation models for territorial species and for nonterritorial species. The final section deals with populations continuously distributed in space, focusing on how spatial environmental autocorrelation and individual dispersal tend to synchronize fluctuations in local populations, and the influence of this synchrony on the risk of regional extinction.

4.2 Metapopulation of a territorial species

Territoriality has been described in a wide variety of animal species, with behavior ranging from the mere occupation of space to the maintenance of interindividual spacing or the patrolling and defense of a home range (Brown, 1975; Davies, 1978). By identifying the unit of suitable habitat as the territory held by an individual female or mated pair, local extinction corresponds to the death of an individual female, and colonization corresponds to individual dispersal and settlement on a suitable, unoccupied territory. This allows the development of a spatially implicit metapopulation model for territorial species that incorporates a general life history as well as the behavior of dispersing individuals searching for a territory. The main purpose of this model is to understand how the equilibrium occupancy of suitable habitat in a large region changes with either habitat destruction and fragmentation, or habitat restoration (Lande, 1987).

Assume for simplicity that a large region is divided into discrete areas the size of individual territories that are either suitable or unsuitable for individual survival and reproduction of a particular species. This model, like Levins', applies to a very large, effectively infinite, region, and is deterministic. Suitable territories are assumed to be randomly or evenly distributed throughout the region, or at least not clumped on a spatial scale greater than the typical individual dispersal distance. Only a single adult female of reproductive age can occupy a suitable territory. The proportion of the region that is suitable habitat for the species is denoted by h. As in classical demography, this model deals with females, assuming no difficulty in mate finding. Let l_τ signify the probability of survival to age τ and f_τ denote the mean number of daughters that survive at least 1 year produced by a female at age τ. The age of adulthood or first reproduction is α. Individuals disperse from their natal areas prior to adulthood, and are assumed capable of searching up to m potential territories before dying if they do not find a suitable unoccupied territory.

Letting p be the proportion of suitable habitat that is occupied by adult females, the probability that a randomly encountered potential territory is either unsuitable or suitable but occupied is $1 - h + ph$. The probability that a dispersing juvenile successfully finds a suitable unoccupied territory is then $1 - (1 - h + ph)^m$. The probability of survival past the age of first reproduction, excluding the risk of dispersal, is signified by l_τ^* so that $l_\tau = l_\tau^*[1 - (1 - h + ph)^m]$ for $\tau \geq \alpha$. At demographic equilibrium, with a constant population size and a stable age distribution, the Euler–Lotka equation (eq. 3.2) with $\lambda = 1$ is

$$[1 - (1 - h + ph)^m]R_0^* = 1 \quad \text{with} \quad R_0^* = \sum_{\tau=\alpha}^{\omega} l_\tau^* f_\tau. \qquad 4.2a$$

Here R_0^* is the mean lifetime production of daughters per newborn female, conditional on the mother finding a suitable unoccupied territory. Note that in this model all of the density dependence occurs during juvenile dispersal in searching for a suitable unoccupied habitat. The equilibrium occupancy of suitable habitat is

$$\hat{p} = \begin{cases} 1 - \dfrac{1-k}{h} & \text{for } h > 1 - k \\ 0 & \text{for } h \leq 1 - k \end{cases} \quad \text{where } k = \left(1 - \dfrac{1}{R_0^*}\right)^{1/m}. \qquad 4.2b$$

The composite parameter k, determined by the life history and dispersal behavior, is called the demographic potential of the population because it gives the maximum occupancy of suitable habitat in a completely suitable region, $\hat{p} = k$ when $h = 1$. Even in a completely suitable region not all of the habitat is occupied at equilibrium because some time generally elapses between a local extinction due to the death of an individual female and the colonization of that territory by a dispersing juvenile. Species with a low demographic potential, either because of a low reproductive output or poor dispersal ability, can occupy only a small fraction of a completely suitable landscape. If newborn females can replace themselves when the population is at low density in a completely suitable region, $R_0^* > 1$, then $0 < k < 1$, and more generally for any h if the equilibrium occupancy is positive then the population will increase when rare and therefore will persist (Lande, 1987).

The model reveals the existence of an 'extinction threshold' or minimum proportion of suitable habitat in a region that is necessary for population persistence, $h = 1 - k$. This is a generic property of spatially implicit metapopulation models for territorial species (Nee, 1994; Hill and Caswell, 1999). As in Levins' model, a population may become extinct in the presence of suitable habitat, but this model specifies the minimum habitat requirement. Figure 4.1 shows how the equilibrium occupancy of suitable habitat changes with the proportion of suitable habitat in the region. For species with a high demographic potential, k near 1, a decrease in the amount of suitable habitat has little effect on the occupancy of suitable habitat until this approaches the extinction threshold where there is a precipitous drop in the equilibrium occupancy. For such species, simply monitoring the occupancy of suitable habitat in response to continuing habitat destruction and fragmentation may give little warning of impending population collapse and extinction. This applies especially when the rate of habitat destruction is rapid compared with the time required for the population

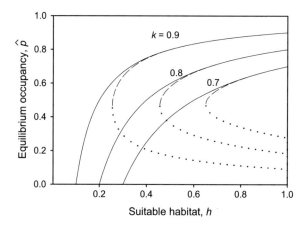

Fig. 4.1 Equilibrium occupancy of suitable habitat, $\hat{p}$, as a function of the proportion of suitable habitat, h, in a large region, for different values of the demographic potential, k, for a territorial species (eq. 4.2b). Solid lines represent stable equilibria assuming no difficulty in mating. Dashed lines represent stable equilibria, and dotted lines represent unstable equilibria, accounting for difficulty in mate finding assuming that dispersing juveniles search $m = 10$ potential territories looking for a suitable unoccupied territory and individuals established on a territory search $\eta = 10$ potential territories seeking a mate (eq. 4.3). Demographic parameters are such that if $\eta = \infty$ (no difficulty in mating) the two models become identical.

to approach demographic equilibrium, as the population will then be above its equilibrium size for the current remaining suitable habitat. Another important conclusion from this model is that with continuing habitat destruction and fragmentation the equilibrium population size declines faster than the amount of suitable habitat, which occurs because the equilibrium population size is proportional to the product of the amount of suitable habitat and the equilibrium occupancy of it. This also may not be apparent under rapid habitat destruction.

This basic model was applied to the Northern Spotted Owl that inhabits the remaining old-growth forests greater than about 250 years old below about 1,200 m elevation in the Pacific Northwest of the U.S.A. Using estimates of amount of old-growth forest remaining in the forested region of the landscape and the occupancy of this habitat by spotted owls, and assuming the population was near a demographic equilibrium, the population was located at a point on the graph in Figure 4.1 (solid lines) which permitted estimation of the demographic potential, $k = 0.79 \pm 0.02$, and the extinction threshold, $h = 1 - k = 0.21 \pm 0.02$. Lande (1988) concluded that the conservation plan developed by the U.S. Forest Service to preserve this subspecies would instead drive it to extinction. Computer simulations of spatially implicit models by Doak (1989) and spatially explicit models by McKelvey *et al.* (1993) and Lamberson *et al.* (1994) demonstrated that clumping of the remaining old-growth habitat would facilitate persistence of a territorial metapopulation by reducing demographic edge effects, that is, the spillover of dispersing juveniles from suitable to unsuitable habitat [as confirmed in general spatially explicit simulations by Bascompte and Solé (1996) and With and King (1999)]. This work, after extensive litigation against the U.S. government, eventually resulted in the development of an improved conservation plan to preserve from commercial logging much of the remaining old-growth forest in numerous large areas in the region, and listing of the subspecies as threatened under the U.S. Endangered Species Act. This model also was applied to estimate the minimum habitat requirement of other avian species (Noon *et al.*, 1997).

The basic metapopulation model for a territorial species can be generalized in several ways (Lande, 1987; Bascompte and Solé, 1996; With and King, 1999; Hill and Caswell, 1999). For example, consider the difficulty of finding a mate in a polygamous mating system, assuming for simplicity that there is a 1:1 sex ratio at conception, with males and females having the same life history and territorial behavior. Further assume that opposite sexes do not exclude each other from suitable territories and that an adult female can mate with any male on her home territory or in η other nearby potential territories. The probability of finding a mate is $1 - (1 - p)(1 - ph)^{\eta}$ and the equilibrium occupancy of suitable habitat by females is obtained by numerical solution of the Euler–Lotka equation at demographic equilibrium (Lande, 1987),

$$[1 - (1 - h + ph)^{m}][1 - (1 - p)(1 - ph)^{\eta}]R_0^* = 1 \qquad 4.3$$

where R_0^* is defined as in equation 4.2a but now also excludes the risk of not finding a mate. Figure 4.1 (dashed lines) reveals a metapopulation Allee effect manifested as an unstable equilibrium at small habitat occupancy, below which the metapopulation is expected to decline to extinction. Comparison of equations 4.1 and 4.3, or the solid and dashed lines in Figure 4.1, shows that the difficulty of mate finding decreases the stable equilibrium occupancy of suitable habitat, and increases the extinction threshold, as also shown by Lamberson *et al.* (1992).

4.3 Metapopulation dynamics of nonterritorial species

We now analyze a nonterritorial species with a population subdivided into discrete spatial locations because of fragmentation of suitable habitat. We derive local extinction and colonization rates from the stochastic local dynamics and explore their effect on metapopulation persistence. Metapopulation dynamics and local population dynamics interact in two ways, both of which operate through changes in habitat occupancy in the metapopulation p because increasing p increases the number of immigrants to each locality. First, increasing p increases the size of local populations and consequently reduces the risk of local extinction, especially for small local populations. This is known as the 'rescue effect' (Brown and Kodric-Brown, 1977), discussed for metapopulations by Hanski (1983) and Hanski and Gyllenberg (1993). Second, increasing p increases the rate of successful colonization, not only by increasing the total rate of dispersal to suitable unoccupied habitat patches, but also by augmenting the probability of successful colonization per migrant through the addition of reinforcements during the crucial initial stages of colony establishment. This is known as the 'establishment effect' (Lande *et al.*, 1998).

4.3.1 Deriving local extinction and colonization rates

Levins' metapopulation model (eq. 4.1) can be extended by allowing the rates of local extinction and colonization to be functions of habitat occupancy of the metapopulation, $e(p)$ and $c(p)$, and to depend on demographic and environmental stochasticity within local populations, assuming that environmental stochasticity operates independently among local populations. Levins' model then becomes

$$\frac{dp}{dt} = c(p)p(1 - p) - e(p)p.$$ 4.4a

Hanski and Gyllenberg (1993) and Gotelli and Kelley (1993) investigated this model by postulating different functional forms for $e(p)$ and $c(p)$, not based on local population dynamics. Below we derive the local extinction and colonization rates from the stochastic dynamics of local populations. These are then employed to analyze a stochastic metapopulation model with a finite number of suitable patches to obtain the mean time to metapopulation extinction (Lande *et al.*, 1998).

As in Levins' model, local populations are viewed as either present or absent on each patch of suitable habitat. We assume that local populations have a positive long-run growth rate below the local carrying capacity, so that the expected duration of local extinction and establishment processes is short compared with the time that local populations spend fluctuating in a quasi-stationary distribution (Chapter 2.6). Denoting the mean local population size on occupied patches as $\bar{N} \equiv \bar{N}(p)$, a successful colonization is defined to occur when a local population founded by a single immigrant to an empty patch of suitable habitat first reaches $\bar{N}$. For each new migrant to an empty suitable patch the probability of successful colonization is written as $u(1, p)$. Assuming a constant rate of individual dispersal or migration between suitable habitat patches, m, such that in an increment of time dt the probability of individual dispersal is $m\,dt$, the expected number of emigrants from an occupied site per unit time is $m\bar{N}$ and the rate of successful colonization is

$$c(p) = m\bar{N}u(1, p). \qquad \text{4.4b}$$

After a successful colonization, the mean time to local extinction is denoted as $T(\bar{N}, p)$. Because the distribution of time to extinction is approximately exponential (Chapter 2.8), the local extinction rate is approximately

$$e(p) = 1/T(\bar{N}, p). \qquad \text{4.4c}$$

Individual dispersal is itself stochastic. With many patches of suitable habitat and a constant dispersal rate per individual, emigration and immigration in each subpopulation can be approximated as independent Poisson processes with variance equal to the mean number of events per unit time (Lande *et al.*, 1998). For concreteness, apart from dispersal between suitable habitat patches, local population dynamics are assumed to have logistic density dependence, $dN/dt = \beta(t)N - \bar{\beta}N/K$, with a constant carrying capacity, K, and stochastic density-independent growth rate, $\beta(t)$, which incorporates any mortality during dispersal, including that caused by dispersal to unsuitable habitat. The diffusion approximation for the size of a local population in a metapopulation with habitat occupancy p then has infinitesimal mean and variance (using the Ito stochastic calculus, Chapter 2.3.3)

$$M(N) = \bar{\beta}N(1 - N/K) - m(N - p\bar{N}) \quad \text{and}$$
$$V(N) = \sigma_e^2 N^2 + \sigma_d^2 N + m(N + p\bar{N}) \qquad \text{4.5a}$$

where σ_e^2 and σ_d^2 are respectively the environmental and demographic variance in $\beta(t)$.

The probability of successful colonization starting from a single individual in a vacant patch of suitable habitat is (Karlin and Taylor, 1981)

$$u(1, p) = \int_0^1 s(N)\,dN \Big/ \int_0^{\bar{N}} s(N)\,dN \qquad \text{4.5b}$$

where $s(N)$ is defined after equation 2.6a. The probability of successful colonization under demographic stochasticity alone was first analyzed in the theory of island biogeography by MacArthur and Wilson (1967) and by Richter-Dyn and Goel (1972).

These authors neglected additional immigrants arriving after the initial colonist, which is a key aspect of the establishment effect.

The mean local population size on occupied patches of suitable habitat, $\bar{N} \equiv \bar{N}(p)$, is the mean of the quasi-stationary distribution at a given habitat occupancy in the metapopulation. This and $T(\bar{N}, p)$, the mean time to local extinction starting from local population size $\bar{N}$ and metapopulation occupancy p, are respectively

$$\bar{N}(p) = \int_0^\infty N \frac{G(N, \bar{N})}{T(\bar{N}, p)} dN \quad \text{and} \quad T(\bar{N}, p) = \int_0^\infty G(N, \bar{N}) dN \qquad 4.5c$$

where the Green function $G(N, \bar{N})$ is defined in equation 2.6a.

Joint numerical solution of formulas 4.5 for any given value of p can be obtained by recursion (the fixed point method usually converges within a few iterations) to find the mean local population size and the local extinction and colonization rates (eqs 4.4b,c), as depicted in Figure 4.2(A). This illustrates that with increasing habitat occupancy in the metapopulation there is a substantial decrease in local extinction rate (the rescue effect) and an even greater increase in colonization rate (the establishment effect). These functions are incorporated into the dynamics of habitat occupancy in the meta-population (eq. 4.4a), as illustrated in Figure 4.2(B). In addition to a stable equilibrium of habitat occupancy, as in Levins' classical metapopulation model, there may be an unstable equilibrium at low habitat occupancy, below which the habitat occupancy is expected to decline to complete metapopulation extinction. This confirms the suggestion of Hanski and Gyllenberg (1993), based on their phenomenological investigation

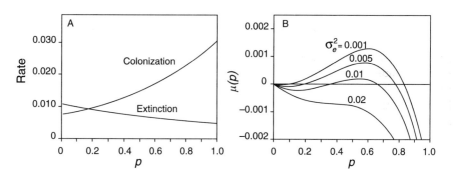

Fig. 4.2 (A) Local extinction and colonization rates as functions of habitat occupancy, p, (eqs 4.4b,c) with a low migration rate. The decrease in local extinction rate and increase in colonization rate with increasing p are, respectively, the rescue effect and the establishment effect. Expected local dynamics are logistic (eq. 4.5a) with deterministic density dependence, carrying capacity $K = 50$, and stochastic density-dependent growth rate with mean intrinsic rate of increase $\bar{\beta} = 0.01$, demographic and environmental variances $\sigma_d^2 = 1$ and $\sigma_e^2 = 0.01$ and expected migration rate $m = 0.005$. (B) Expected rate of change of habitat occupancy, $\mu(p)$, as a function of p, with parameter values as in panel A, for different environmental variances σ_e^2. (From Lande *et al.*, 1998.)

of equation 4.4a, that an Allee effect can occur at the metapopulation level for non-territorial species. Evidently, a certain minimum habitat occupancy can allow the rescue and establishment effects to maintain a stable metapopulation, when a geographically isolated local population would not persist for long due to a combination of demographic and environmental stochasticity and emigration.

Many species of vertebrates satisfy the usual assumption of a constant rate of individual dispersal (Gaines and McClenaghan, 1980). However, in some species local population density affects the rate of emigration (Wolff, 1997). Positive density dependence of juvenile dispersal rate has been observed in several species of birds and mammals, and negative density-dependent emigration has been recorded in species that may tend to aggregate for social reasons (Wolff, 1997; Doncaster *et al.*, 1997; Diffendorfer *et al.*, 1999). Population density is often confounded with habitat quality and individual habitat selection behavior during dispersal, which complicates the notion of population 'sources and sinks' (Pulliam, 1988; Pulliam and Danielson, 1992) as discussed by Doncaster *et al.* (1997). The number of conspecifics in an area may be utilized by (potentially) dispersing individuals as an indicator of habitat quality (Kiester, 1979; Danchin *et al.*, 1991; Stamps, 1991). However, high population density can degrade habitat quality. For example, in colonial seabird species, immigration rate often decreases with increasing population size due to difficulty in obtaining a favorable breeding site (Croxall and Rothery, 1991).

The impact of density-dependent dispersal on the rescue and establishment effects was analyzed by Sæther *et al.* (1999). In a metapopulation model similar to that in equations 4.4 and 4.5, they showed that positive density dependence of emigration and negative density dependence of immigration rates tend to magnify the rescue and establishment effects that increase the stable equilibrium occupancy of habitat in a metapopulation.

4.3.2 Mean time to extinction of a finite metapopulation

With independent environmental stochasticity among local populations, at a given metapopulation habitat occupancy, p, local extinction and colonization become Poisson processes at the level of the metapopulation. A finite metapopulation on a limited number of suitable habitat patches is therefore subject to stochastic extinction. With a total of n patches of suitable habitat, if i habitat patches are occupied (successfully colonized) then $p = i/n$. In an infinitesimal time interval dt denote the probability of an increase from i to $i + 1$ occupied patches as $B(i)dt$ and the probability of a decrease from i to $i - 1$ occupied patches as $D(i)dt$, where

$$B(i) = c(p)p(1 - p) \quad \text{and} \quad D(i) = e(p)p. \qquad 4.6a$$

There is a reflecting barrier at full habitat occupancy, $i = n$, so that $B(n) = 0$. This is formally equivalent to a density-dependent birth-and-death process (Goel and Richter-Dyn, 1974; Leigh, 1981; Nisbet and Gurney, 1982), for which the mean

Table 4.1 Mean times to metapopulation extinction, starting at
the stable equilibrium occupancy of suitable habitat, depending
on the number of suitable habitat patches, n, and the individual
dispersal rate between them, m (from eqs 4.4, 4.5, 4.6). Expected
local dynamics are logistic with $K = 50$, $\bar{\beta} = 0.01$, $\sigma_e^2 = 0.01$,
$\sigma_d^2 = 1$

m	n	This model[1]	Classical model[2]
0.005	50	3.7×10^3	9.3×10^{10}
	100	6.5×10^3	8.3×10^{19}
0.01	50	1.9×10^{17}	4.1×10^{52}
	100	2.5×10^{32}	1.8×10^{105}

[1] With $e(p)$ and $c(p)$ evaluated using equations 4.5.
[2] With constant e and c evaluated at the stable equilibrium occupancy.

time to metapopulation extinction starting from i occupied patches is given by the
recursion

$$T(i+1) = T(i) + \sum_{j=i+1}^{n} \frac{B(i) \cdots B(j-1)}{D(i) \cdots D(j-1)} \frac{1}{D(j)} \qquad 4.6b$$

where $T(0) = 0$ and by definition $B(0)/D(0) = 1$.

Mean times to metapopulation extinction incorporating the rescue and establishment effects are compared in Table 4.1 with those in a classical metapopulation model
with constant rates of local extinction and colonization. Both models start from the
stable equilibrium occupancy of suitable habitat in the metapopulation, with a rate of
individual dispersal between suitable habitat patches, m. For example, in this model
accounting for rescue and establishment effects, with $m = 0.005$ and $\sigma_e^2 = 0.01$,
there is a stable equilibrium habitat occupancy of the expected dynamics at $\hat{p} = 0.65$,
as shown in Figure 4.2. With constant extinction and colonization rates corresponding to the stable equilibrium habitat occupancy in the metapopulation, the classical
model produces extremely large overestimates of the persistence of the metapopulation, because it fails to account for an increasing local extinction rate and decreasing
colonization rate as habitat occupancy declines, the rescue and establishment effects
(Fig. 4.2A). The classical model also can greatly underestimate the mean time to
metapopulation extinction if local extinction and colonization rates used in the model
are those for a single isolated local population (Lande *et al.*, 1998).

Accurate models of metapopulation dynamics therefore should incorporate the
stochastic dynamics of local populations and the rescue and establishment effects. In
practice, local populations often differ in their expected dynamics, and metapopulation properties also may depend on the spatial distribution of suitable habitat that
may vary in quality and be subject to spatial environmental autocorrelation (Hastings
and Harrison, 1994; Harrison and Bruna, 1999). For example, the invasion of a new
habitat by an exotic species frequently begins in a single locality and occurs in a wave

encompassing an expanding geographic range (Elton, 1958; Shigesada and Kawasaki, 1997). The rate of spread often accelerates after the initial stages, which may be caused by evolutionary adaptation to a new environment, by chance introduction to a relatively unfavorable locality, or by an Allee effect or demographic stochasticity (Lewis and Kareiva, 1993; Veit and Lewis, 1996; Hastings, 1996; Crooks and Soulé, 1996). The later stages of spread also may be stochastic because of rare long-distance dispersal and habitat heterogeneity (Shigesada and Kawasaki, 1997; Suarez *et al.*, 2001; Keeling *et al.*, 2001).

Realistic simulations of spatially explicit models generally are so complex that their results are best interpreted qualitatively, e.g. by comparing the relative rather than absolute effects of alternative management actions (McKelvey *et al.*, 1993; Dunning *et al.*, 1995; Kareiva and Wennergren, 1995; Wennergren *et al.*, 1995; Lindenmayer *et al.*, 2000, 2001). Analysis of simple models is nevertheless useful to clarify the basic mechanisms operating in more complex situations that can only be studied by simulation. Section 4.4 analyzes the influence of spatial environmental autocorrelation.

4.4 Population synchrony

The correlation of temporal fluctuations in population size between localities is called population synchrony. Populations can become synchronized by common environmental factors such as weather (Grenfell *et al.*, 1998) or predation (Stenseth *et al.*, 1999; Ims and Andreassen, 2000). Individual dispersal between populations also can act to synchronize them. Closely related species that occupy the same region but differ in dispersal ability tend to differ in the spatial scale over which local populations are synchronized (Hanski and Woiwod, 1993; Sutcliffe *et al.*, 1996; Lindström *et al.*, 1996; Myers *et al.*, 1997*a*; Ranta *et al.*, 1998; Koenig, 1998; Paradis *et al.*, 2000). Population synchrony in a given species also may differ among regions (Stenseth *et al.*, 1999; Paradis *et al.*, 2000). Patterns of population synchrony observed in butterflies and fish are depicted schematically in Figure 4.3.

Raw data on population synchrony are quite noisy because of sampling error in population estimates and limited time spans of observation, as shown in Fig. 4.4. Sampling error in population estimates tend to impart a downward bias in observed population synchrony. This occurs because sampling variance inflates estimated variances in population size in the denominator of a correlation coefficient, but does not alter the expected covariance between independently estimated populations in the numerator of a correlation coefficient. Estimated patterns of population synchrony therefore do not approach the theoretical correlation of unity at zero distance (as for the butterfly data in Fig. 4.3) unless they are constrained to do so by curve fitting techniques (as for the fish data in Fig. 4.3).

Substantial effort has been made to understand the relative contributions of dispersal versus spatial environmental autocorrelation to observed patterns of population

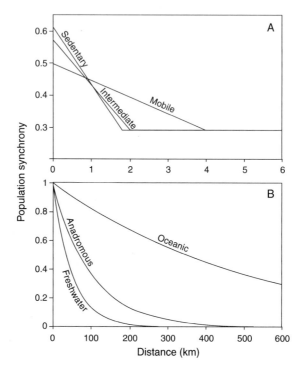

Fig. 4.3 Schematic diagrams of empirical spatial patterns of population synchrony. (A) Spatial patterns of population synchrony for butterfly species with sedentary, intermediate, and mobile dispersal. Modified from Sutcliffe *et al.* (1996). (B) Spatial patterns of synchrony in recruitment for oceanic, anadromous and freshwater fish species. From data of Myers *et al.* (1997a). (From Lande *et al.*, 1999.)

synchrony. Moran (1953) analyzed a simple linear model of two closed populations subject to correlated environmental stochasticity, which would apply to small fluctuations around an equilibrium with no individual dispersal. He concluded that under identical forms of expected dynamics in both populations, with no individual dispersal between them, the correlation between populations equals the environmental correlation. This synchronization of population fluctuations by environmental autocorrelation is called the 'Moran effect.'

Computer simulations of nonlinear population dynamics suggest that spatial environmental autocorrelation contributes more to population synchrony than does individual dispersal, especially over long distances (Ranta *et al.*, 1997, 1998). Simulations also reveal that population synchrony strongly influences regional population dynamics and the risk of regional or global extinction (Allen *et al.*, 1993; Bolker and Grenfell, 1996; Heino *et al.*, 1997; Ranta *et al.*, 1997; Palmqvist and Lundberg, 1998; Kendall *et al.*, 2000). However, no general quantitative relationships between population synchrony and the spatial scales of environmental autocorrelation,

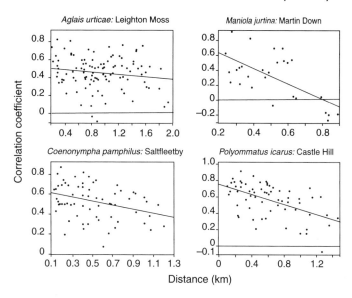

Fig. 4.4 Raw autocorrelation data between all pairs of local populations within four butterfly species monitored over at least 10 years as a function of geographic distance. (From Sutcliffe *et al.*, 1996.)

individual dispersal, and other factors emerged from the simulations. This happened partly because in nonlinear models with no individual dispersal, large environmental fluctuations tend to reduce population synchrony in comparison with the spatial environmental correlation, diminishing the Moran effect (Grenfell *et al.*, 1998; Greenman and Benton, 2001). Here we develop an analytical theory of small or moderate fluctuations for spatially distributed populations in which the expected dynamics are homogeneous in space, with individual dispersal and spatially autocorrelated environmental stochasticity (Lande *et al.*, 1999). We also describe a generalization to large fluctuations in population size, including the risk of regional extinction (Engen *et al.*, 2002*b*).

4.4.1 Stochastic dynamics with dispersal

Assume that the habitat is an infinite two-dimensional space with geographic coordinates z_1 and z_2. The population density at location $z = (z_1, z_2)$ at time t is denoted as $N \equiv N(z, t)$. Local population sizes are assumed large enough to neglect demographic stochasticity. The expected population dynamics are spatially homogeneous with no age structure and no time delays. Individual dispersal is isotropic such that from every location the distribution of individual displacement vectors, $y = (y_1, y_2)$, is symmetric, $f(y) = f(-y)$, with a mean displacement of 0. The individual dispersal rate is m so that in an increment of time dt the probability of individual dispersal is mdt. A general model of population growth and dispersal in continuous time and space is

$$\frac{dN(z,t)}{dt} = \beta(z, N, t)N(z, t) - mN(z, t) + m \int N(z - y, t)f(y)\, dy. \quad 4.7$$

The per capita rate of population growth at each location, $\beta(z, N, t)$, includes density dependence as well as stochasticity. The second term is the total emigration rate from locality z and the last term is a convolution of the functions $mN(z, t)$ and $f(y)$ describing the total rate of immigration to locality z averaged over all possible source locations, $z - y$, weighted by the probability of individual dispersal to the locality. A constant risk of individual mortality during dispersal would effectively reduce m and $\beta(z, N, t)$ without changing the general form of the model.

The distribution of individual dispersal displacement vectors in equation 4.7 is more accurate, especially over short distances and small times, than the commonly used diffusion approximation for individual movement as a continual Brownian motion in space. Observed distributions of individual dispersal distance per lifetime, $R = \sqrt{y_1^2 + y_2^2}$, usually are strongly skewed with a long tail, representing a high proportion of short-distance movements and a small proportion of long-distance movements (Kot *et al.*, 1996; Veit and Lewis, 1996; Turchin, 1998), as illustrated in Figure 4.5. If individual dispersal is observed in an area then the one-dimensional distribution of dispersal distances is the radial integral of the two-dimensional distribution of dispersal displacement vectors. For example, with a radially symmetric distribution of dispersal vectors $f(y_1, y_2)$, changing from Euclidean to radial coordinates

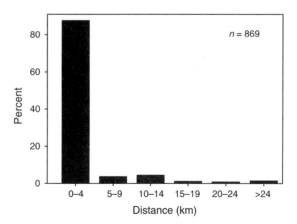

Fig. 4.5 Natal dispersal in the House Sparrow in a 65 km square study area on islands in northern Norway where a large proportion of all individuals are color-ringed (Ringsby *et al.*, 1999). Search for color-ringed individuals was also conducted outside the main study area. There is no sexual difference in the distribution of dispersal distance.

involves the Jacobean factor R, that is $f(y_1, y_2)dy_1dy_2 = \mathsf{R}\, f(\mathsf{R})\,d\mathsf{R}\,d\theta$ so that the one-dimensional distribution of dispersal distances is $\int_0^{2\pi} \mathsf{R}f(\mathsf{R})d\theta = 2\pi\mathsf{R}f(\mathsf{R})$.

4.4.2 Linearized model of small fluctuations

For small or moderate fluctuations in local population size equation 4.7 can be linearized around the stable deterministic equilibrium or carrying capacity, K. Linearized models are accurate within about 10% when coefficients of variation for fluctuations in local population size are as large as 30% (Lande *et al.*, 1999), which is within the range often observed in actual populations (Table 1.1). Environmental stochasticity may occur in the density-independent growth rate and/or carrying capacity of local populations, which are assumed to remain sufficiently large to neglect demographic stochasticity. The proportional deviation from the deterministic equilibrium, $\varepsilon(z, t) = N(z, t)/K - 1$, with $\mathrm{E}[\varepsilon(z, t)] = 0$ at each location, obeys the stochastic integrodifferential equation (see Chapter 2)

$$d\varepsilon(z, t) = -(\gamma + m)\varepsilon(z, t)dt + mdt \int \varepsilon(z - x, t)f(x)\,dx + \sigma_e dB(z, t). \quad 4.8$$

Here γ is rate of return to equilibrium under small perturbations in the absence of dispersal (see Chapter 3). Local population growth rate is subject to environmental stochasticity with no temporal autocorrelation, described by the increment of spatially autocorrelated Brownian motion, $dB(z, t)$ (Chapter 2), with environmental standard deviation σ_e.

Because equation 4.8 is linear, small fluctuations in relative size of local populations are normally distributed in space and time. Thus their stationary distribution can be described by the spatial and temporal variances and covariances of local fluctuations. The stationary coefficient of covariation (the squared coefficient of variation times the spatial correlation of local population size) between local populations separated by a displacement vector y is denoted as $c(y) = \mathrm{E}[\varepsilon(z, t)\varepsilon(z + y, t)]$. The squared coefficient of variation of the stationary distribution of local population size is $c(0)$. We denote the spatial correlation function between local populations, that is the population synchrony, as $\rho(y) = c(y)/c(0)$. Environmental stochasticity is assumed to be spatially (but not temporally) autocorrelated, such that its correlation between localities depends only on the displacement, y, between them, $\mathrm{E}[dB(z, t)dB(z + y, t)] = \rho_e(y)dt$ (Chapter 2.3.3) where $\rho_e(y)$ denotes the spatial environmental autocorrelation function.

To derive a formula for $c(y)$ write $\varepsilon(z, t + dt) = \varepsilon(z, t) + d\varepsilon(z, t)$ where $d\varepsilon(z, t)$ is given by equation 4.8, and write a similar formula for $\varepsilon(z + y, t + dt)$. Taking the expectation of their product at the stationary distribution of population fluctuations, ignoring terms of order dt^2, and cancelling a factor of $c(y)$ from both sides, yields

$$2(\gamma + m)c(y) = 2m \int c(y - x)f(x)\,dx + \sigma_e^2 \rho_e(y). \quad 4.9a$$

This is a Fredholm integral equation for $c(y)$, which has a unique solution (Korn and Korn, 1968, p. 497) that can be derived using Fourier transforms, and that is

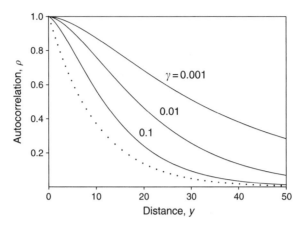

Fig. 4.6 Spatial autocorrelation function for different strengths of density regulation γ around carrying capacity K. The dotted line is the spatial autocorrelation of environmental stochasticity, corresponding formally to $\gamma = \infty$. The spatial scale of individual dispersal is $l = 2$ with dispersal rate $m = 1$ and spatial scale of environmental autocorrelation $l_e = 10$. (Modified from Lande *et al.*, 1999.)

useful for numerical analysis and graphics such as Figure 4.6 (Lande *et al.*, 1999). An elementary method of solution is to treat equation 4.9a as a recursion. Writing $c_{j+1}(y)$ on the left side and $c_j(y)$ on the right side instead of $c(y)$, starting from $c_0(y) = 0$, the solution for the stationary coefficient of covariation, $c(y) \equiv c_\infty(y)$, is

$$c(y) = \frac{\sigma_e^2}{2(\gamma + m)} \int \rho_e(y - x) \sum_{i=0}^{\infty} \left(\frac{m}{\gamma + m} \right)^i f^{i*}(x)\, dx \qquad 4.9b$$

where $f^{i*}(y) = \int f(y - x) f^{(i-1)*}(x)\, dx$ is the i-fold convolution of the dispersal distribution with itself, and $f^{0*}(x) = \delta(x)$, the Dirac delta function, which is a spike with infinite height, infinitesimal width, and unit area centered $x = 0$, with the property $\int f(y - x)\delta(x)\, dx = f(y)$ (Korn and Korn, 1968). Thus the stationary coefficient of covariation is a convolution of the environmental autocorrelation function with a weighted sum of all possible i-step dispersal events.

The coefficient of variation of local population size is bounded by the values for no dispersal and long-distance dispersal. In the absence of dispersal ($m = 0$) at equilibrium $c(y) = \sigma_e^2 \rho_e(y)/(2\gamma)$, and thus the synchrony between local populations is the same as the environmental correlation, $\rho(y) = \rho_e(y)$. This exemplifies the 'Moran effect' showing that, without dispersal, spatially correlated environmental stochasticity acts to synchronize population fluctuations (Moran, 1953). At the opposite extreme, with long-distance dispersal $f(x)$ becomes broad and flat as most individual dispersal distances approach infinity, so that only the first term in the summation in equation 4.9b gives a contribution and $c(y) \to \sigma_e^2 \rho_e(y)/[2(\gamma + m)]$. Then synchrony between local populations again equals the spatial environmental autocorrelation, as with no dispersal. This occurs because, in the limit of infinitely long-distance dispersal, immigrants to each locality are drawn from such a wide area that the source populations are effectively independent and the rate of immigration to each locality approaches a constant that cannot contribute to population synchrony. The squared coefficient of variation therefore has the bounds

$$\frac{\sigma_e^2}{2(\gamma + m)} < c(0) < \frac{\sigma_e^2}{2\gamma}. \qquad 4.10$$

Thus for dispersal to greatly decrease the coefficient of variation it is necessary for the dispersal rate to exceed the strength of density dependence, $m > \gamma$.

In contrast to the extreme cases of no dispersal and long-distance dispersal, which do not alter population synchrony, local dispersal can substantially increase population synchrony. Numerical examples of the stationary pattern of population synchrony as a function of the distance between local populations are depicted in Figure 4.6, for particular forms of the distribution of dispersal displacements and the spatial environmental autocorrelation function. Because the bivariate population autocorrelation function describing population synchrony is radially symmetric about the origin, it can be illustrated as a univariate function by interpreting y as the distance from the origin in a particular direction, say along the y_1 axis. Figure 4.6 suggests that under weak density dependence the spatial scale of population synchrony can substantially exceed the spatial scale of environmental autocorrelation, even when individual dispersal is short compared with the scale of environmental autocorrelation. To ascertain the generality of this conclusion we analyze the spatial scale of population synchrony in relation to the spatial scales of environmental autocorrelation and individual dispersal.

4.4.3 Spatial scale of population synchrony

The scale of a distribution or function can be measured in different ways. A commonly used measure is the distance over which a function decays to $e^{-1} \cong 0.37$ of its maximum value, which is especially useful for exponential functions. For example, the radially symmetric distribution of dispersal distance $f(y_1, y_2) = e^{-R/l}/(2\pi l^2)$ has a spatial scale l. Another measure of scale often used to characterize the width of a probability distribution is the standard deviation. This measure can be applied to any function that can be normalized to integrate to unity. It allows the scale of a two-dimensional function to differ in different directions. For example, the squared spatial scale of a general individual dispersal distribution along the y_1 axis is the variance $l^2 = \iint y_1^2 f(y_1, y_2) \, dy_1 \, dy_2$.

Employing the standard deviation as a measure of scale produces a simple, general formula for the spatial scale of population synchrony in terms of the spatial scales of dispersal and environmental autocorrelation. The spatial scales for population synchrony and environmental autocorrelation, denoted respectively as l_ρ and l_e, can be defined as the standard deviations of the corresponding functions normalized to integrate to unity. Assuming that dispersal and spatial environmental autocorrelation are local (vanishing at long distances), population synchrony also will be local and the variances defining these scales will be finite. The stationary solution for $c(y)$ (eq. 4.9b) shows that if a regional component of spatial environmental autocorrelation exists it produces a regional component of population synchrony, that is, a constant background correlation as in Figure 4.3(A), which causes the scales l_e and l_ρ to become infinite. This defect can be repaired by subtracting the constant background level from the covariance functions $c(y)$ and $\rho_e(y)$ in the linear equation 4.9a, thereby restricting the formulas to only local covariance with finite scales.

The spatial scale of population synchrony can be derived from the stationary solution for $c(y)$, but a simpler approach is to obtain it directly from equation 4.9a as follows. Divide each term in equation 4.9a by $I = \int c(y)\,dy$ and take the variance along the y_1 axis by multiplying both sides by y_1^2 and integrating over space. The distribution of a sum of two independent random variables is the convolution of their distributions, so the variance of the convolution is the sum of the variances of the two variables (Stuart and Ord, 1994). This yields $2(\gamma + m)l_\rho^2 = 2m(l_\rho^2 + l^2) + \sigma_e^2(J/I)l_e^2$ where $J = \int \rho_e(y)\,dy$. Integrating equation 4.9a over space establishes that $\sigma_e^2 J/I = 2\gamma$ since $I^{-1}\iint c(y - x)f(x)\,dx\,dy = 1$, which holds because the convolution of two normalized distributions also is a normalized distribution that must integrate to 1 (Stuart and Ord, 1994). These two relations finally produce (Lande *et al.*, 1999)

$$l_\rho^2 = l_e^2 + \frac{ml^2}{\gamma}. \qquad 4.11$$

This simple relationship shows that relative to the scale of spatial environmental autocorrelation the contribution of dispersal to the spatial scale of population synchrony is magnified by the ratio of the dispersal rate to the strength of density regulation, m/γ. For dispersal to be the major factor contributing to population synchrony requires that $ml^2/\gamma > l_e^2$. This may be true even when the scale of dispersal is less than the scale of spatial environmental autocorrelation, $l < l_e$, if the strength of density dependence is much less than the dispersal rate, $\gamma \ll m$, as illustrated in Figure 4.6. Weak density dependence permits large fluctuations in local population size (eq. 4.10), which can then spread by local dispersal to neighboring localities before they are damped by density dependence, thereby augmenting the spatial scale of synchrony among local populations. For species with very weak density dependence the scale of population synchrony becomes very large. As $\gamma \to 0$ equation 4.11 implies that $l_\rho \to \infty$, analogous to a critical point in a physical phase transition (Stanley, 1987). In simple models without age structure the strength of density dependence is proportional to the density-independent growth rate (see after eqs 5.1). Thus with a long-run growth rate below carrying capacity, $\bar{r}$, near zero, as for many threatened and endangered species, local extinctions will tend to occur synchronously over large regions, insofar as habitat fragmentation has not interrupted individual dispersal among local populations. Of course, with permanent habitat heterogeneity, local populations in the most consistently favorable areas will tend to persist longer (Hastings and Harrison, 1994; Harrison and Bruna, 1999).

The same simple scaling relationship is valid for the log of local population size under large fluctuations subject to Gompertz density dependence (eqs 1.9) because the dynamics remain linear, provided that individual dispersal obeys a continuous Brownian motion in space (Engen, 2001). Similar scaling laws also apply with permanent spatial heterogeneity in the environment modeled as a stationary spatial process (Engen et al., 2002a).

4.4.4 Synchrony in annual population change

The coefficient of covariation of populations displaced in space by the vector y and in time by lag τ is defined as $c(y, \tau) = E[\varepsilon(z, t)\varepsilon(z + y, t + \tau)]$. From equation 4.8 a derivation analogous to that for formulas 4.9 produces the differential equation

$$\frac{dc(y, \tau)}{d\tau} = -(\gamma + m)c(y, \tau) + 2m \int c(y - x)f(x)\,dx \qquad 4.12a$$

in which $c(y) \equiv c(y, 0)$ is the stationary solution (eq. 4.9b). Equation 4.12a can be solved using Fourier transforms (Lande et al., 1999)

$$c(y, \tau) = e^{-\gamma \tau} \int c(y - x) \sum_{j=0}^{\infty} e^{-m\tau} \frac{(m\tau)^j}{j!} f^{j*}(x)\,dx. \qquad 4.12b$$

From this the squared spatial scale of population synchrony at time lag τ can be derived as $l_\rho^2 = l_e^2 + (1/\gamma + \tau)ml^2$, which is larger than that at time lag 0 by τml^2, the variance of dispersal distance in time τ.

The spatial scale of synchrony in annual changes in local population size has been estimated by some authors (Bjørnstad et al., 1999a,b; Viljugein et al., 2001). Assuming that most of the changes in local population size are caused by local birth and death rather than dispersal, the synchrony in annual population changes should closely approximate the environmental autocorrelation function. The synchrony of small annual changes in population size, $\text{Cov}[\Delta N(z, t), \Delta N(z+y, t)]$, is proportional to $c(y) - c(y, 1)$. The same applies for log population size. From equation 4.12b it can be shown that in the absence of dispersal, $m = 0$, the synchrony of annual population change is the same as the spatial environmental autocorrelation. Thus the Moran effect for the synchrony of population size also applies to population change. In general, with dispersal, the spatial scale of synchrony in annual population change, l_{diff}, is the same as that for $c(y) - c(y, 1)$, which can be approximated using the continuous time formulas 4.11 and 4.12b as

$$l_{\text{diff}}^2 = l_e^2 + \frac{ml^2}{\gamma}\left(1 - \frac{\gamma}{e^\gamma - 1}\right) = l_e^2 + \frac{ml^2}{2}\left(1 - \frac{\gamma}{6} + \cdots\right). \qquad 4.13$$

(Engen et al., 2002b). If the scale of environmental autocorrelation is much larger than the scale of individual dispersal, $l_e^2 \gg ml^2/2$, then the scale of synchrony in

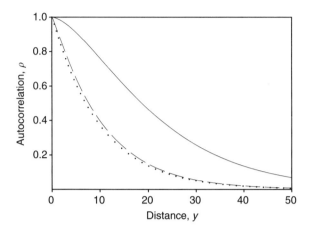

Fig. 4.7 Spatial autocorrelation function for population size (solid line), for environmental stochasticity (dotted line) and for annual difference in abundance (dashed line) with $\gamma = 0.01$. Other parameters as in Figure 4.6.

annual population changes (or annual changes in log population size) accurately represents the spatial scale of environmental autocorrelation, $l_{\mathrm{diff}} \cong l_e$. This situation is illustrated in Figure 4.7 with a numerical example of spatial autocorrelation functions for environmental stochasticity, population size, and annual population change.

4.4.5 Local and regional extinction risk

The risk of regional extinction depends primarily on the magnitude of local population fluctuations and their synchrony resulting from the net influence of density dependence, spatial environmental autocorrelation, and individual dispersal. Decreasing the strength of density dependence increases the magnitude of local population fluctuations (eq. 4.10) and increases the spatial scale of population synchrony (eq. 4.11), which increase local and regional extinction risks, respectively. Increasing the spatial scale of environmental autocorrelation increases the spatial scale of population synchrony (eq. 4.11), which increases regional extinction risk.

Dispersal increases the rate of return of local fluctuations to the mean local population size and reduces the coefficient of variation of local population size (eq. 4.10). With large fluctuations and Gompertz density regulation, dispersal also increases the local mean population size (Engen *et al.*, 2002*b*). Each of these effects decreases the risk of local and regional extinction. On the other hand, dispersal increases the spatial scale of population synchrony (eq. 4.11), which increases the risk of regional extinction as neighboring local populations tend to fluctuate and become extinct together. Dispersal thus exerts opposing influences on the risk of regional extinction, but the overall effect of dispersal is to decrease local extinction risk as illustrated in Figure 4.8.

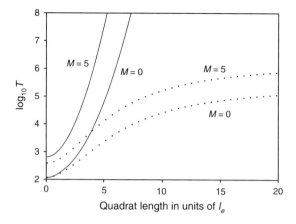

Fig. 4.8 The logarithm of the expected time to quasi-extinction (at 10% of the local carrying capacity K) in relation to quadrat length for different values of the variance of individual dispersal distance in a given direction per unit time, M. Solid lines are for purely local spatial autocorrelation of environmental stochasticity, with $\rho_e(\infty) = 0$. Dotted lines are for both local and global components in the spatial environmental autocorrelation, with $\rho_e(\infty) = 0.25$. The local component of the environmental autocorrelation function has a Gaussian form with spatial scale $l_e = 10$. Other parameters are $\sigma_e^2 = 0.10$ and $\gamma = 0.04$. The spatial scale of (the local component of) population synchrony is $l_\rho = l_e$ for $M = 0$ and $l_\rho = 1.5 l_e$ for $M = 5$ (from eq. 4.11 replacing ml^2 with M).

Regional extinctions tend to occur on a spatial scale commensurate with that of synchrony between local populations. Figure 4.8 shows that the mean time to regional quasi-extinction increases sharply in quadrats with side lengths larger than the scale of population synchrony (Engen *et al.*, 2002*b*). The size of protected areas for species conservation, therefore, ideally should be substantially larger than the scale of population synchrony.

4.5 Summary

Metapopulation models incorporating stochastic local dynamics have been developed for territorial and for nonterritorial species. For territorial species local extinction can be identified with the death of an individual resident female, and colonization occurs with the successful dispersal of an individual female searching for a suitable unoccupied territory. Territorial species subject to random habitat fragmentation have a minimum amount of suitable habitat in a region necessary for metapopulation persistence, termed the extinction threshold. Under continued habitat fragmentation the equilibrium occupancy of suitable habitat declines faster than the amount of suitable habitat, until the extinction threshold is reached. For nonterritorial species, local extinction and colonization rates can be analyzed using diffusion approximations incorporating demographic and environmental stochasticity and individual dispersal. This reveals that with increasing habitat occupancy the local extinction rate decreases and the colonization rate increases, due to the rescue and establishment effects. These effects can cause a metapopulation to be far less viable than in a classical meta-population model that assumes constant rates of local extinction and colonization. Metapopulation persistence may depend on the existence of currently unoccupied suitable (or temporarily unsuitable) habitat.

For populations continuously distributed in space, with uniform expected dynam-ics, we derive a linearized model of small or moderate fluctuations in local population size. We analyze how the spatial autocorrelation of local population size (population synchrony) depends on the pattern of spatial environmental autocorrelation, the dis-tribution of individual dispersal distances, and the strength of density dependence. Defining the scale of a probability distribution or normalized function by its standard deviation produces an especially simple scaling relationship among these processes. This shows that under weak density dependence the spatial scale of population syn-chrony may substantially exceed the spatial scale of environmental autocorrelation, even with a relatively small scale of individual dispersal. In contrast, the spatial scale of synchrony in annual population changes usually is close to the scale of spatial environmental autocorrelation. Regional extinctions tend to occur on a spatial scale similar to that of population synchrony. The size of areas for species conservation should therefore be substantially larger than the spatial scale of population synchrony.

5

Population viability analysis

5.1 Assessing extinction risk

5.1.1 Qualitative assessment criteria

Species differ greatly in abundance. In diverse communities, such as lowland tropical birds, the distribution of individual abundance among species at a given site often is approximately lognormal, with a substantial number of species represented by only a few individuals even in very large samples (Preston, 1948, 1962, 1980; Williams, 1964). Thus many species are naturally rare at a given site. Human disturbance of natural communities is detrimental to most species, benefiting relatively few species, and therefore usually reduces the species diversity of ecological communities (Patrick, 1963; Robbins *et al.*, 1989; Azuma *et al.*, 1997; Lande *et al.*, 2000). In many developed and developing areas of the world, human disturbance has extinguished numerous species and has caused a substantial proportion of the remaining species to become threatened or endangered with extinction. The major anthropogenic factors threatening species are, in order of decreasing importance: (1) habitat destruction and fragmentation, primarily for agriculture and human habitation; (2) overexploitation by commercial and subsistence hunting for food and the live animal pet trade; (3) introduced species of competitors, predators and parasites; and (4) pollution, which threatens species mainly near point sources but can cause widespread morbidity and mortality (Groombridge, 1992; Heywood, 1995; Hilton-Taylor, 2000).

The World Conservation Union (or International Union for Conservation of Nature and Natural Resources, IUCN) developed objective, population-based criteria for assessing the extinction risk of animal and plant species (IUCN, 2001). For each of the risk categories Critically Endangered, Endangered, and Vulnerable, there are several criteria concerning population size, fragmentation, trends, and fluctuation, as well as geographic range and area occupied, that determine the level of risk assigned to the species. A capsule summary of these criteria are presented in Table 5.1. Most species of animals and plants that are reasonably well known to science, especially vertebrates and flowering plants, and some groups of insects such as butterflies, can be categorized as to their extinction risk using the IUCN criteria by panels of experts on the taxa. Species at risk of extinction according to these criteria are compiled in the IUCN Red Lists for each country or region of the world and updated periodically. The proportions of mammal and bird species of the world that appear on the IUCN Red Lists are compiled in Table 5.2.

Table 5.1 Summary of IUCN (2001) Red List categories and criteria. N is population size and T is generation time. For times given in years or generations, the longer one applies up to a maximum of 100 years. In criteria B and C the amount of population fragmentation depends on the category, and fluctuation refers to extreme fluctuation of population size by a factor of 10.

Criterion	Category		
	Critically Endangered	Endangered	Vulnerable
A. ΔN in 10 yr or $3T$	$\Delta N < -90\%$, causes known and ceased OR $\Delta N < -80\%$, causes unknown or continuing	$\Delta N < -70\%$, causes known and ceased OR $\Delta N < -50\%$, causes unknown or continuing	$\Delta N < -50\%$, causes known and ceased OR $\Delta N < -30\%$, causes unknown or continuing
B. Area	Range < 100 km² OR occupancy <10 km² AND 2 of fragmentation, continuing decline, fluctuation	Range < 5000 km² OR occupancy <500 km² AND 2 of fragmentation, continuing decline, fluctuation	Range < 20,000 km² OR occupancy <2000 km² AND 2 of fragmentation, continuing decline, fluctuation
C. N and ΔN	$N < 250$ AND continuing decline with $\Delta N < -25\%$ in 3 yr or $1T$ OR fragmentation or fluctuation	$N < 2500$ AND continuing decline with either $\Delta N < -20\%$ in 5 yr or $2T$ OR fragmentation or fluctuation	$N < 10,000$ AND continuing decline with either $\Delta N < -10\%$ n 10 yr or $3T$ OR fragmentation or fluctuation
D. N	$N < 50$	$N < 250$	$N < 1000$ OR occupancy <20 km² or <5 locations
E. PVA	Extinction risk > 50% in 10 yr or $3T$	Extinction risk >20% in 20 yr or $5T$	Extinction risk >10% in 100 yr

Table 5.2 Numbers of all known mammal and bird species of the world in IUCN extinction risk categories (Hilton-Taylor, 2000 using 1994 IUCN criteria)

	Total species	Critically Endangered	Endangered	Vulnerable	% threatened	% near threatened
Mammals	4763	180	340	610	23.7%	12.6%
Birds	9946	182	321	680	11.9%	7.3%

Alarming proportions of the world's mammal and bird species are threatened with extinction during the next 100 years, and additional substantial proportions are near threatened. The total numbers of threatened species of herpetofauna (reptiles and amphibians), fish, and invertebrates are comparable with those for mammals and birds (Hilton-Taylor, 2000). Other taxa are not as thoroughly known as birds and mammals, and all of their known species have not been assessed by IUCN, so it is not possible to accurately estimate the percentage of total species in other taxa that are threatened.

The figures in Table 5.2 may actually underestimate the magnitude of the problem because the IUCN criteria do not account for the long-term effects of global climate change caused by human emission of greenhouse gases, which can interact with and exacerbate other anthropogenic factors contributing to extinction risk. For example, theory and paleontological evidence suggest that change of geographic range has been a major mechanism promoting population persistence in response to past climate changes, allowing populations to track the habitats to which they are adapted as the habitats move across the landscape (Coope, 1979; Pease et al., 1989; Peters and Lovejoy, 1992; Smith et al., 1995; McCarty, 2001). Large-scale anthropogenic alteration of the landscape and resulting habitat fragmentation may inhibit this process in the future by restricting individual dispersal and reducing the opportunity for species to alter their geographic ranges.

Simple descriptors of species life history and ecology, such as body size, generation time, trophic level, habitat or dietary specialization, individual home range size, and species geographic range, do not provide good predictors of extinction risk for species on the IUCN Red Lists. Although certain variables of this kind are significantly associated with extinction risk in particular higher taxa, they do not singly or in combination have much predictive power across higher taxa because the factors contributing most to extinction risk differ among species and higher taxa (Purvis et al., 2000). Quantitative assessment of extinction risk for species or populations must therefore be based on detailed population models.

5.1.2 Quantitative assessment by population viability analysis (PVA)

For a small minority of species, sufficient information exists on their life history, demography, and ecology to permit the construction and analysis of an accurate population model and quantitative assessment of extinction risk. This endeavor is known as PVA. Because of the substantial proportions and high numbers of threatened species in many ecosystems, and the lack of data to perform PVAs on the great majority of species, increasing emphasis is being placed on large-scale conservation and restoration of landscapes and ecosystems (Edwards *et al.*, 1994; Scott *et al.*, 1996; Soulé and Terborg, 1999). However, large-scale conservation can succeed only if it preserves much of the species diversity in the planning area. PVA for selected species of special interest or ecological importance can help to validate large-scale conservation plans. PVA should therefore be an integral part of large-scale conservation efforts.

Modern concepts of PVA were initiated when Shaffer (1981) pointed out that quantitative assessment of extinction risk should be based on a stochastic model of population dynamics, with population viability defined by the probability of extinction within a certain time, e.g. a probability of extinction less than 10% in 100 years or less than 1% in 1,000 years. According to Shaffer (1981) a stochastic model for assessing population viability would ideally incorporate not only demographic and environmental stochasticity but also genetic stochasticity, including inbreeding and random genetic drift, which can affect individual fitness through inbreeding depression and loss of genetic variability reducing the potential for evolutionary adaptation to changing environments (Franklin, 1980; Soulé, 1980; Gilpin and Soulé, 1986; Lande, 1995). Shaffer (1983) performed the first PVA incorporating only demographic stochasticity in a model of a small population of Brown Bears. Since then many PVAs have been done on a variety of animal and plant species, most of them using generic computer programs for age- or stage-structured populations such as VORTEX (Lacy, 1993), RAMAS (Ferson and Akçakaya, 1990), and ALEX (Possingham and Davies, 1995). For explication and reviews see Boyce (1992), Burgman *et al.* (1993), Lindenmayer *et al.* (1995), Schemske *et al.* (1994), Groom and Pascual (1998), and Beissinger and Westphal (1998).

PVA is one of the cornerstones of conservation biology, and has stimulated research on factors affecting extinction risk for particular species in addition to theoretical work on stochastic population modeling (Sjögren-Gulve and Ebenhard, 2000; Beissinger and McCullough, 2002). Sensitivity analysis of population models can be useful in deciding which aspects of the life history or ecology require additional studies to improve model accuracy, and also can help to decide which management actions are likely to be most effective in improving mean population growth rate and reducing extinction risk (McCarthy *et al.*, 1995; Hedrick *et al.*, 1996; Caswell, 2001). When converted to population density, PVA can be used to determine the minimum size of suitable habitat necessary to support a viable population (Armbruster and Lande, 1993; Beier, 1993).

5.1.3 Biases and uncertainties in PVAs

PVA has recently received much criticism. PVA is included as one of the principal criteria for each of the primary threat categories in the IUCN Red Lists (Table 5.1), but it is almost never employed in the Red Lists, because PVAs are considered unreliable. As emphasized by many authors, PVAs often are subject to serious statistical biases and uncertainties (Taylor, 1995; Mills *et al.*, 1996; Ludwig, 1996*b*, 1999; Beissinger and Westphal, 1998; Engen and Sæther, 2000; Fieberg and Ellner, 2000; Coulson *et al.*, 2001*b*; Sæther and Engen, 2002).

In practice, PVAs are based primarily on demographic and environmental stochasticity, sometimes including potential catastrophes. Models of genetic stochasticity affecting population viability, if incorporated at all, generally are not very realistic due to lack of detailed information on complex evolutionary processes affecting fitness, including adaptive evolution, random genetic drift, inbreeding, and gene flow (Lande, 2002). Below we consider the main weaknesses of demographic models in PVA and suggest a general statistical method for incorporating uncertainty in population parameters using prediction intervals for population size.

Population sizes and trends often are estimated with substantial uncertainty and bias, especially for rare and endangered species (Solow 1998; Taylor and Wade, 2000; Holmes, 2001; Holmes and Fagan, 2002). Even when population estimates have high accuracy, estimates of population parameters, such as mean population growth rate and carrying capacity, or age- or stage-specific vital rates, usually have substantial uncertainty for three reasons. First, population size and age or stage structure typically are only estimated rather than completely counted. Second, the limited duration of most population studies, combined with demographic and environmental stochasticity makes it unlikely that the years of observation precisely estimate the long-term average and variability of population parameters. Third, the initial age or stage distribution in the population is often unknown and therefore not taken into account in PVAs, which can bias population projections (Chapter 3; Taylor, 1995).

Another common bias in PVAs is that demographic stochasticity often is not separated from environmental stochasticity, thus inflating estimates of the environmental variance. Chapter 1 describes methods for estimating demographic variance and environmental variance and separating these from density dependence, using data on individual fitness within years and precise time series observations of population sizes. Sampling variance in population estimates, if known from mark–recapture studies (White *et al.*, 2002) or repeated estimates of population sizes, can be subtracted from estimates of the environmental variance. In age- or stage-structured populations, estimation of environmental variances and covariance in vital rates requires time series analysis of vital rates and/or population age or stage structure to separate effects of density dependence, which has seldom been done (e.g. Benton *et al.*, 1995; Caswell, 2001). Because of the limited duration of most population studies, rare extreme environmental fluctuations, including catastrophes, are either poorly estimated or completely absent from the observations, leading to underestimation of environmental stochasticity (Ludwig, 1999; Fieberg and Ellner, 2000; Coulson *et al.*, 2001*b*).

The strength and form of density dependence is often unknown or estimated with low precision, especially in age- or stage-structured models. The form of density dependence can have a substantial influence on estimates of extinction probabilities (Mills *et al.*, 1996; Middleton and Nisbet, 1997; Sæther *et al.*, 2000*a*). A solution to this problem is to use a general form of density dependence that encompasses a variety of simple types of density dependence. Incorporating time series data on environmental covariates, such as temperature, precipitation, and/or abundance of strongly interacting species, can help explain observed time series of population fluctuations and increase the statistical precision of population parameters (Caughley and Gunn, 1993; Pech and Hood, 1998; Davis *et al.*, 2002) and population viability analysis (Dennis and Otten, 2000). However, it is impossible to estimate accurately the expected population dynamics outside the range of observed population sizes, for example a possible Allee effect at very small population sizes (Myers *et al.*, 1995).

A general conclusion for density-dependent populations, assuming a positive long-run growth rate below carrying capacity and a stationary distribution of environments, is that the distribution of time to extinction begins with a period of low extinction risk, while the population size approaches a quasi-stationary distribution, after which the extinction risk of surviving populations becomes nearly constant per unit time. The initial period of low extinction risk is apparent even for density-independent populations with a negative long-run growth rate (eq. 2.11 and Figs 2.12, 2.13), because some time must elapse before the population declines to extinction. This tends to produce a deceptively optimistic view of population viability in the short term. For this reason (Armbruster *et al.*, 1999) recommended that, especially for species with long generations, such as elephants, extinction risk should be assessed over a range of time frames, including some much longer than a few generations.

PVAs of declining populations often assume a negative long-run growth rate with no density dependence, implying that the population is doomed to relatively rapid extinction (Dennis *et al.*, 1991; Gaston and Nicholls, 1995; Holmes and Fagan, 2002). In many cases declining populations may instead be tracking a declining carrying capacity due to continuing habitat degradation, which implies that the population decline could be halted if habitat degradation ceases and the population achieves a new density-dependent equilibrium (e.g. Ratner *et al.*, 1997). PVAs assuming density-independent declines typically include environmental stochasticity but ignore demographic stochasticity (Dennis *et al.*, 1991; Gaston and Nicholls, 1995; Holmes and Fagan, 2002). Neglecting demographic stochasticity fails to account for its increasing importance in smaller populations (Chapter 1), and the consequent acceleration of the final decline to extinction (Fig. 2.4; Engen and Sæther, 2000), which can substantially underestimate extinction risk (e.g. Fig. 5.7).

Declining populations with a negative long-run growth rate should not be considered viable simply because they have less than 10% probability of extinction within 100 years, if they have a substantial probability of extinction within a somewhat longer time such as 200 years, particularly for species with long generations (Armbruster *et al.*, 1999). Even for species with a positive long-run growth rate below carrying capacity, the initial period of low extinction risk in distributions of time to extinction (Chapter 2), suggests that PVAs should consider a range of time frames rather than only those times dictated by convention. PVA predictions are likely to be most accurate on relatively short time scales (Beissinger and Westphal, 1998; Brook *et al.*, 2000; Engen and Sæther, 2000; Fieberg and Ellner, 2000; Sæther *et al.*, 2000a, Engen *et al.*, 2001), but may also be accurate over longer time scales for species with high adult annual survival rates and long generations so that stochasticity and uncertainty accumulate slowly with time.

Estimation of dispersal in spatially distributed populations is especially problematic. Dispersing individuals may suffer high mortality, making it difficult to observe successful dispersal. Individual dispersal behavior may be extremely complex, depending on the habitat quality and population density in source and potential settlement locations and on the arrangement and quality of intervening habitat in the landscape (Chepko-Sade and Halpin, 1987; Doncaster *et al.*, 1997; Diffendorfer *et al.*, 1999; Lindenmayer and Lacy, 2002). Long-distance dispersal is often missed, leading to truncated distributions of dispersal distance and underestimation of dispersal rates.

PVAs tend to be overly optimistic in assessments of population viability for several reasons. Risk factors are often underestimated, especially extreme environmental events, Allee effects, global climate change, and synergistic interactions of various demographic and genetic risk factors. PVAs usually fail to account fully for the impacts of future increases in human population (Lacy and Miller, 2002). While attempting to account for stochasticity, PVAs often ignore or underestimate extinction risk due to uncertainty in population sizes and population parameters. The limited time frame of most PVAs implies that they generally overestimate long-term population viability. However, accurate predictions of extinction risk cannot generally be made over time frames much longer than the period of observed population fluctuations (Fieberg and Ellner, 2000).

A conceptual bias pervades most PVAs, which focus more on stochastic population dynamics *per se* than on the ecological factors that determine population parameters. Caughley (1994) emphasized that PVAs typically shed little light on the deterministic causes of population declines, which can only be ascertained by painstaking field studies of all potentially relevant ecological factors (see case studies in Caughley and Gunn, 1996).

5.2 Incorporating uncertainty using population prediction intervals (PPIs)

Despite the limitations of PVA, quantification of extinction risk requires some type of PVA based on a stochastic population model. With sufficient data, PVA is potentially more objective and informative than qualitative methods of classifying extinction risk that fail to specify the probability of extinction within a certain time. An essential goal of PVA should be to incorporate the uncertainty arising from limited data, as well as stochasticity in population dynamics, in forecasting extinction risk. This section develops techniques that fully incorporate uncertainty in population parameters into PVA predictions. We begin by briefly considering the role of uncertainty in applications of the Precautionary Principle to natural resource management.

According to the Precautionary Principle (O'Riordan *et al.*, 2001) advocated by IUCN and many conservationists (IUCN, 2001), uncertainty should favor listing rather than not listing a species as being threatened with extinction. In view of the substantial proportions of species threatened with extinction (Table 5.2), the burden of proof should rest on those who claim a species is not threatened, rather than vice versa. Uncertainty due to limited data therefore should be incorporated in PVA by increasing the chance that a species is listed. Because of difficulties in accurately estimating Allee effects at small population sizes, and the frequent lack of data on individual fitness necessary to estimate demographic stochasticity, PVAs are increasingly often performed by simulating stochastic population declines not to extinction but instead to a specified level that is sufficiently large to neglect Allee effects, demographic stochasticity (see after eq. 1.3), and inbreeding depression in fitness. The use of such a quasi-extinction threshold not only improves the accuracy of PVAs by reducing uncertainty in population forecasts, but also provides a precautionary early warning of situations where a species has an appreciable chance of becoming endangered. The limited data available on most threatened and endangered species nevertheless imposes substantial uncertainties on estimates of population parameters that should be incorporated in a precautionary fashion in PVAs.

Uncertainties in population parameters in PVAs have been analyzed using sensitivity analysis of extinction probabilities to changes in population parameters, usually one at a time. Even simultaneous perturbations of multiple population parameters does not fully account for joint distribution of uncertainties in all population parameters in the estimation of extinction probabilities.

Several authors have advocated Bayesian approaches to PVA to include all uncertainties in population parameters (Ludwig, 1996b; Taylor *et al.*, 1996; Lee, 2000; Goodman, 2002; Wade, 2002). Various shortcomings of Bayesian methods, as well as its allure, are lucidly discussed by Dennis (1996). A particular drawback of Bayesian methods and decision theory, as applied to natural resource management and PVA, is that they may encourage the substitution of human beliefs for biological data, rendering these methods susceptible to subjective bias and inhibition of data collection.

Here we describe and apply the general method of PPI to incorporate stochasticity in population dynamics and uncertainty in population parameters into PVA. The concept of prediction intervals is fundamental in statistics (McCullagh and Nelder, 1989; Box *et al.*, 1994; Stuart and Ord, 1999). Prediction intervals concern the likely range of an unobserved variable, in this case future population size. In contrast, confidence intervals concern the likely range of an observed or estimated parameter. Heyde and Cohen (1985) and Dennis *et al.* (1991) previously employed prediction intervals for density-independent populations subject to environmental stochasticity, but neglected demographic stochasticity and assumed no uncertainty in population parameters. We generalize this approach to include density dependence, demographic stochasticity, and uncertainty in parameter estimates.

A PPI is defined as the stochastic interval that includes the unknown population size at a specified future time with a given probability or confidence level. A PPI is stochastic because it is based on data on past stochastic population fluctuations that are used to estimate the population parameters (see Chapter 1). A PPI at a given time t in the future is constructed using a large number of simulations of a stochastic population model over t years. A random set of population parameters is used in each simulation, drawn from the sampling distribution of the parameters, e.g. obtained by bootstrapping (Efron and Tibshirani, 1993; Sæther *et al.*, 2000*a*). The statistical accuracy (or coverage) of a PPI can be checked and adjusted by additional simulations for known parameter values (Engen and Sæther, 2000; Engen *et al.*, 2001).

In PVA it is most appropriate to use upper one-sided prediction intervals for population size, as the main interest concerns whether the population size at any given time is above a certain size or whether the population will become extinct before a given time. If t_C is the shortest time in which the PPI includes the (quasi)extinction boundary at $N = C$, then we predict (quasi)extinction to occur after time t_C with the confidence level used to construct the PPI. The width of the PPI increases with increasing stochasticity and increasing uncertainty. Uncertainty in population parameters does not change the extinction risk of the population, but it does affect our confidence in population predictions and the risk of (quasi)extinction.

For known population parameters, the IUCN category of Vulnerable is defined according to Criterion E by a 10% probability of extinction within 100 years (Table 5.1). The natural generalization of this to unknown parameters is to categorize a species as Vulnerable if the upper 90% prediction interval for population size at 100 years includes extinction. In the special case of known parameters the shortest time at which the upper 90% PPI includes extinction will correspond to the time at which the cumulative probability of extinction is 10% (see Chapter 2.8), showing that the PPI is a proper generalization of the IUCN criterion to the case of unknown parameters. The PPI is consistent with the Precautionary Principle because uncertainty due to limited data expands the PPI to include (quasi)extinction in a shorter time, thus increasing the chance that the species will be listed, and requiring more data to exclude a species from being listed.

Our analysis is based on simple stochastic population models without age structure, although the method can be extended to age-structured populations (Heyde and Cohen, 1985). Insufficient data exists for most threatened species to support complex age-structured models when fully accounting for uncertainty in all the vital rates. The greater total uncertainty in models with a large number of parameters would tend to produce wider prediction intervals. It is well established in statistics that relatively simple models with a small or moderate number parameters provide more predictive power (Burnham and Anderson, 1998). Simple population models incorporating estimates of net demographic and environmental stochasticity, and appropriate density dependence, may produce sufficiently realistic descriptions of the dynamics of many threatened and endangered species to obtain meaningful results from PVA see eqs 3.10a,b. For simple life histories, like that in Chapter 3.1.4, it may prove feasible to estimate the vital rates and to account fully for uncertainty in them to calculate a PPI.

A PPI incorporates uncertainty due to limited data, producing a statement about the likely range of population size, and whether this includes (quasi-)extinction within a specified time. It is easily interpreted and communicated to managers and decision makers. In contrast, the alternative approach of using a confidence interval for the probability of extinction within a specified time produces a statement about the probability of the probability of extinction within a certain time, e.g. there is at least a 10% chance that the probability of extinction is greater than 10% in 100 years (Ludwig, 1999; Wade, 2002). This is not easily interpreted in terms of the population size itself and is therefore likely to confuse population managers and decision makers. PPIs thus have the advantage of simplicity of interpretation. When the population parameters are known exactly the distribution of time to extinction, and the cumulative probability of extinction before a given time, can be computed exactly and contain all the information about extinction risk (Chapter 2.8). However, with uncertainty in parameter estimates there is no necessary reason why we should focus primarily on the unknown probability of extinction. We should instead be concerned with statistical inferences about the population size and the time of extinction, which is the purpose of prediction intervals. They express our confidence in the actual population size and extinction time, rather than our confidence in the probability of extinction within a specified time.

5.3 Population models

5.3.1 Density regulated populations

A general form of density regulation that includes the logistic and Gompertz models as special cases is provided by the theta-logistic model (Gilpin and Ayala, 1973). A discrete-time stochastic version of the theta-logistic model with stochastic density independent growth and deterministic density dependence is

$$\frac{\Delta N}{N} = \bar{\beta}\left[1 - \left(\frac{N}{K}\right)^{\theta}\right] + \varepsilon(t) \qquad 5.1a$$

where $\varepsilon(t)$ is the stochastic component of the density-independent growth rate with mean 0 and variance $\sigma_e^2 + \sigma_d^2/N$ where σ_e^2 and σ_d^2 are, respectively, the environmental and demographic variances (Chapter 1). Different values of θ produce a range of forms of density dependence. For $\theta = 1$ density dependence takes the linear, logistic form, and $\theta = \infty$ describes density-independent growth to a ceiling (Table 2.3, Fig. 2.8). The dynamics can be rewritten for $\theta \neq 0$ as

$$\frac{\Delta N}{N} = \bar{\beta}_1\left[1 - \frac{N^{\theta} - 1}{K^{\theta} - 1}\right] + \varepsilon(t) \qquad 5.1b$$

where $\bar{\beta}_1 = (1 - K^{-\theta})\bar{\beta}$ is the expected per capita growth rate at a population size of $N = 1$. This form clarifies that in the limit as $\theta \to 0$ the density-dependent (fractional) term takes the Gompertz form, $-\ln N/\ln K$ (Diserud and Engen, 2000). We employ the form of the dynamics in equation 5.1 and define (quasi)extinction to occur at a population size $N = 1$. Figure 5.1(A) compares the expected dynamics for different values of θ keeping $\bar{\beta}_1$ constant. The corresponding strength of density dependence per year as a function of population size, $\gamma = -\partial \ln \lambda/\partial \ln N$ (Chapter 3.4.1), is depicted in Figure 5.1(B).

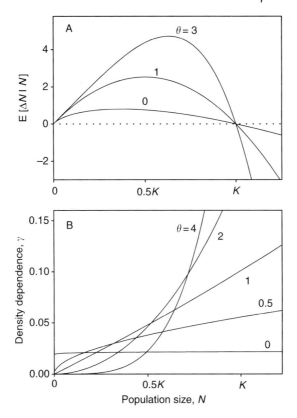

Fig. 5.1 (A) Expected dynamics as a function of population size in the theta-logistic model, for different values of θ, with $\bar{\beta}_1 = 0.1$ and $K = 100$. (B) Strength of density dependence per year, $\gamma = -\partial \ln \lambda / \partial \ln N$, as a function of population size, N, for different values of θ in the theta-logistic model (eq. 5.1).

For populations with a small coefficient of variation, which remain near carrying capacity, the dynamics are described by a linearized version of this model. Assuming that both $N - K$ and $\varepsilon(t)$ are small quantities, expanding equation 5.1a around the deterministic equilibrium at $N = K$ and retaining only first-order terms yields the linearized equation

$$\Delta N = -\theta \bar{\beta}(N - K) + K \varepsilon(t)$$

in which the expected rate of return to equilibrium is $\gamma = \theta \bar{\beta}$. This reveals that for a density regulated population with a small coefficient of variation it is not possible to estimate accurately both θ and $\bar{\beta}$ because these only appear as their product γ, which can be accurately estimated. Estimates of θ and $\bar{\beta}$ therefore tend have a negative sampling correlation, for populations with stable expected dynamics, $-1 < \gamma < 1$.

5.3.2 Declining populations

In the absence of continued habitat degradation, or based on evidence that density dependence is absent or weak, populations with a clear declining trend can be assumed to be density independent with a negative long-run growth rate. For such populations we can employ the stochastic density-independent model

$$\frac{\Delta N}{N} = \bar{\lambda} - 1 + \varepsilon(t) \qquad\qquad 5.2a$$

where $\varepsilon(t)$ has mean 0 and variance $\sigma_e^2 + \sigma_d^2/N$. A nearly equivalent model describes the dynamics directly on the natural log scale, $X = \ln N$, with

$$\Delta X = \bar{r} + \varepsilon(t) \qquad\qquad 5.2b$$

where $\bar{r}$ is the long-run growth rate and again $\varepsilon(t)$ has mean 0 and variance $\sigma_e^2 + \sigma_d^2/N$ (see Chapters 1 and 2).

5.4 Examples of PVA

5.4.1 Song Sparrow on Mandarte Island

During 24 years of complete censuses the Song Sparrow breeding population on Mandarte Island, B.C., Canada fluctuated between 4 and 72 pairs, as shown in Figure 5.2. The demographic variance was estimated from individual fitnesses as the weighted mean demographic variance within years, $\hat{\sigma}_d^2 = 0.66$. The remaining population parameters in the stochastic theta-logistic model (eq. 5.1b) were estimated from the time series in Figure 5.2 by the method in Chapter 1, assuming no

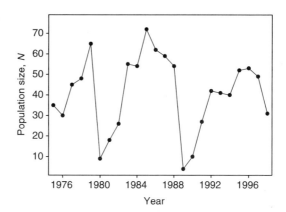

Fig. 5.2 Time series of the breeding population size for the Song Sparrow on Mandarte Island, B.C., Canada. (From Sæther *et al.*, 2000*a*.)

environmental autocorrelation (Sæther *et al.*, 2000*a*). The estimated form of density dependence, $\hat{\theta} = 1.09$, indicates that the expected dynamics are nearly logistic. The expected density-independent growth rate and the carrying capacity were respectively estimated as $\hat{\beta}_1 = 0.99$ and $\hat{K} = 41.5$. The estimated environmental stochasticity was $\hat{\sigma}_e^2 = 0.41$.

The demographic variance is estimated with relatively high precision, based on extensive individual data over the duration of the study. Virtually all of the uncertainty occurs in the other parameters. Sampling distributions of these parameters were obtained by computer simulations of the population time series using the estimated parameters in the discrete-time stochastic recursion (eq. 5.1b), and re-estimating the parameters from each simulation assuming that the demographic variance is known. This produced the sampling distributions in Figure 5.3, which reveal substantial uncertainty in the parameter estimates, particularly $\hat{\theta}$. In addition there was a strong negative sampling correlation between $\hat{\beta}_1$ and $\hat{\theta}$.

The PPI was calculated at each time by stochastic simulations (Efron and Tibshirani, 1993; Sæther *et al.*, 2000*a*). The population dynamics were simulated using each

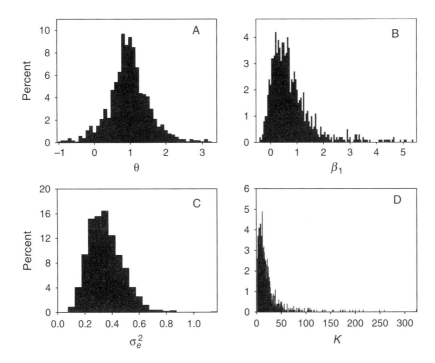

Fig. 5.3 Uncertainty in parameter estimates for the Song Sparrow on Mandarte Island. Graphs show the marginal distributions each parameter obtained from parametric bootstrapping by simulation of the population model using the bias-corrected parameter estimates. (Modified from Sæther *et al.*, 2000*a*.)

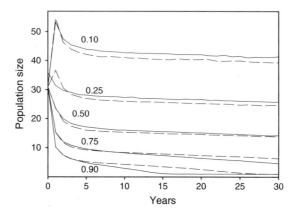

Fig. 5.4 Population Prediction Interval (PPI) for Song Sparrow on Mandarte Island. Solid lines give the lower boundaries of the upper specified portions of the PPI in future years, accounting for demographic and environmental stochasticity and uncertainty in the joint sampling distribution of the parameters (Fig. 5.3). Dashed lines give the corresponding boundaries of the PPI accounting for stochasticity, but assuming no uncertainty in the parameters. (Modified from Sæther *et al.*, 2000*a*.)

bootstrap replicate of the parameter values. This method accounts for sampling correlations in the parameters. The lower boundaries of upper PPIs for different confidence values are illustrated in Figure 5.4. Also depicted are the corresponding lower bounds of the PPIs based only on demographic and environmental stochasticity, ignoring uncertainty in population parameters, assuming that the parameters are known and equal to their estimates. Comparison of the PPIs with and without uncertainty reveals that accounting for uncertainty in parameter estimates substantially increases the breadth of the PPIs. For any given confidence level, accounting for uncertainty produces a more precautionary assessment of extinction risk, indicated by the shortest time at which the PPI includes the extinction boundary at $N = 1$. For example, using the upper 90% confidence value for the PPI, the shortest time at which the PPI includes the extinction boundary is 17 years, whereas ignoring uncertainty this level of extinction risk is not reached until 30 years. Thus including uncertainty substantially reduces the time for the extinction boundary to be included in the PPI.

5.4.2 Barn Swallow in Denmark

The migratory Barn Swallow has declined substantially in many areas of Europe in part because of conversion to modern agricultural practices that reduce reproductive success by diminishing insect prey during the breeding season (Møller, 2001). Nearly complete censuses of nesting pairs in a farmland area at Kraghede, Denmark showed a declining trend from 184 pairs to 59 pairs from 1984 to 1999 (Fig. 5.5). Møller (1989) detected density dependence in reproduction, but not in winter survival, during an initial non-declining period 1971–1979, and was unable to detect density

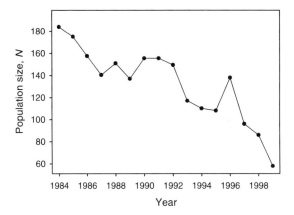

Fig. 5.5 Time series of the number of breeding pairs of the Barn Swallow at Kraghede, Denmark. (From Engen *et al.*, 2001.)

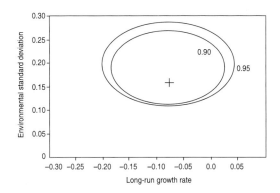

Fig. 5.6 Uncertainty in parameter estimates for the Barn Swallow at Kraghede. The + indicates the bias-corrected estimates of the environmental standard deviation, σ_e, and the long-run growth rate, $\bar{r}$, and the inner and outer ellipses represent respectively the 90% and 95% confidence regions for the joint sampling distribution. Skewness of the joint distribution with respect to σ_e arises from the Chi-squared sampling distribution of the environmental variance. (Modified from Engen *et al.*, 2001.)

dependence during the later declining period 1980–1988. It is therefore reasonable to suppose that since 1984 the population summering at Kraghede has undergone a density-independent decline. Although the population is not closed to immigration and emigration, offspring usually return to their natal areas to breed. To illustrate the application of the model of a declining density-independent population (eq. 5.2), we analyzed data from this population to estimate the risk of local extinction within a range of times, excluding immigration from other populations.

Extensive data on individual fitness within years and the 16-year population time series in Figure 5.5 were analyzed using the method in Chapter 1 to estimate the population parameters in equation 5.2b. The weighted mean demographic variance within years was estimated as $\hat{\sigma}_d^2 = 0.18$. Estimates of the environmental variance and the long-run growth rate were $\hat{\sigma}_e^2 = 0.024$ and $\bar{r} = -0.076$. The PPI was calculated by computer simulations of the time series as described above, accounting for uncertainty in $\hat{\sigma}_e^2$ and $\bar{r}$, but assuming that the demographic variance is known (Engen *et al.*, 2001). Lower boundaries for the upper PPIs corresponding to different confidence levels are shown in Figure 5.6. Ignoring uncertainty in the parameters, assuming they were known and equal to their estimates, the upper 90% confidence value for the PPI includes extinction in 28 years (Fig. 5.6). Accounting for uncertainty in the parameter estimates (Fig. 5.5) decreases this time substantially to 21 years. This illustrates again that accounting for uncertainty results in a more precautionary assessment of extinction risk.

PVAs of declining populations typically ignore demographic stochasticity (Dennis *et al.*, 1991; Gaston and Nicholls, 1995; Holmes, 2001). Assuming the demographic variance is zero causes a constant (approximating the average value of σ_d^2/N in the population time series) to be added to the estimated environmental variance. This inflates the environmental variance and misrepresents the increase in stochasticity with population declines produced by demographic stochasticity, which may seriously underestimate the extinction risk, as shown in Figure 5.7. Ignoring the demographic variance, the lower boundary of the upper 90% confidence value for the PPI includes extinction in 33 years, instead of 21 years when demographic and environmental variance are separated.

5.5 Summary

For most species insufficient information exists to construct an accurate quantitative population model, but qualitative assessment of extinction risk can be performed using objective, population-based systems such as the IUCN Red List criteria. IUCN lists substantial proportions of animal taxa as threatened with extinction during the next 100 years. Quantitative PVA uses a stochastic population model to estimate the probability of extinction of a population or species before a certain time. PVAs have been criticized for several reasons, including sensitivity to different forms of population dynamics, low predictive power due to statistical uncertainty in population parameters, and for analyzing population viability at a single time in the future. These criticisms can be addressed by the following procedures.

(1) Use a population model, such as the theta-logistic, with sufficient flexibility to encompass a wide range of forms of density dependence. PVAs of declining populations that assume a negative long-run growth rate with no density dependence should provide ecological evidence to support these assumptions versus the alternative of density dependence with a declining carrying capacity.

(2) Set a quasi-extinction threshold at a population size large enough to neglect demographic stochasticity, Allee effects, and genetic stochasticity. This will reduce

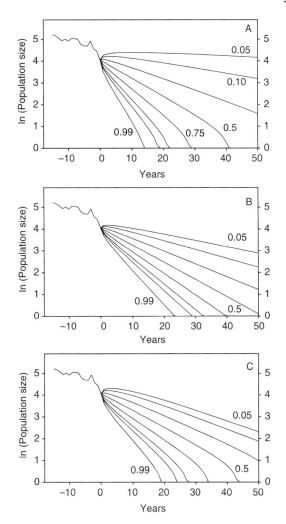

Fig. 5.7 Population Prediction Interval (PPI) for the Barn Swallow at Kraghede. Lines give lower boundaries of the upper specified portions of the PPI in future years (A) accounting for demographic and environmental stochasticity and uncertainty in the joint distribution of the parameters (Fig. 5.6), (B) assuming no uncertainty in the parameters, and (C) assuming no demographic stochasticity, $\sigma_d^2 = 0$, so that all stochasticity is estimated as environmental stochasticity. (Modified from Engen *et al.*, 2001.)

uncertainty and help provide an early warning of when a population is likely to become threatened.

(3) Incorporate statistical uncertainty in population parameters using PPIs over a range of times.

A PPI is the stochastic interval that includes the unknown population size at a specified future time with a given probability or confidence. The concept of population viability can be generalized to incorporate uncertainty and the Precautionary Principle, by employing the upper PPI with a specified probability. For example, the IUCN category of Vulnerable is defined by Criterion E as a 10% probability of extinction within 100 years. With uncertainty this naturally generalizes to categorizing a species as Vulnerable if the upper 90% PPI at 100 years includes extinction. As a direct statement about a future population size, a PPI is easier to interpret and to communicate to population managers and decision makers than the alternative approach of estimating a confidence interval for the probability of extinction at a future time.

6

Sustainable harvesting

6.1 Overexploitation and its causes

Natural populations often are exploited for subsistence or commerce. Rapidly increasing resource demands from growing human populations and modern harvesting technology, coupled with inadequate management and regulation, have caused overexploitation of many species to the point of population collapse, sometimes to near extinction or to actual extinction. About half of the commercial fisheries in the U.S.A. and Europe were recently classified as overexploited (Rosenberg et al., 1993). Over one-third of the endangered mammals and birds of the world are threatened with extinction in part by hunting and collecting, and one-eighth are threatened by international trade, including traffic in live animals for the pet trade (Groombridge, 1992; Redford, 1992; Hilton-Taylor, 2000). Commercial harvesting has driven several species to near extinction, including the Blue Whale *Balaenoptera musculus*, Right Whale *Eubalaena* spp., Northern Elephant Seal *Mirounga angustirostris*, American Bison *Bison bison*, Black Rhinoceros *Diceros bicornis*, White Rhinoceros *Ceratotherium simum*, White Abalone *Haliotis sorenseni*, and Barndoor Skate *Raja levis* (Gulland, 1971; LeBoeuf, 1981; Cumming et al., 1990; Miller, 1990; Ludwig et al., 1993; Primack, 1993; Casey and Myers, 1998; Davis et al., 1998). Overexploitation combined with habitat destruction caused local extinctions or severe depletion of many species, including the African Elephant *Loxodonta africana*, Jaguar *Panthera onca*, Tiger *Panthera tigris*, and numerous commercial fisheries (Cumming et al., 1990; Redford, 1992; Linden, 1994; Hutchings, 2000). Some of these species have partially recovered following cessation of legal hunting, but others may become extinct in the foreseeable future because of continued habitat destruction, poaching, and/or Allee effects. Commercial harvesting or subsistence hunting were major factors in the extinction of several species, including the Great Auk *Pinguinus impennis* (extinct 1844), Passenger Pigeon *Ectopistes migratorius* (perhaps the most abundant terrestrial bird species that ever existed, extinct 1914), Steller's Sea Cow *Hydrodamalis stelleri* (extinct 1768), and Caribbean Monk Seal *Monachus tropicalis* (extinct 1962) (Halliday, 1980; Groombridge, 1992).

Overexploitation impacts not only the harvested species but also frequently has detrimental impacts on other species. Fisheries often exert heavy mortality on non-target species, referred to as bycatch (Alverson et al., 1994; Pope et al., 2000). For example, decline and virtual extinction of the Barndoor Skate occurred as a result of bycatch mortality in the northwest Atlantic trawl fishery (Casey and Myers, 1998).

Long-line fisheries cause high bycatch mortality in albatrosses and petrels in the North Pacific and Southern Ocean; drift-net and gill net fisheries kill large numbers of diving birds such as puffins (Croxall *et al.*, 1997; Tasker *et al.*, 2000). Fisheries that have been severely overexploited often failed to recover at anticipated rates when targeted fishing pressure has been curtailed, in part because of bycatch mortality due to fishing other relatively abundant species (Hutchings, 2000). Exploitation of coastal marine communities has often followed an historical pattern of initial depletion of some species, with compensatory increases in competing species, followed by their overexploitation, finally leading to functional collapse of the community (Lichatowich, 1999; Jackson *et al.*, 2001). Overexploitation of target species can cause cascading ecological effects. For example, exploitation of sea urchin predators, including Sea Otters *Enhydra lutris*, large fish, and Spiny Lobsters *Panulirus* spp., leads to destructive overgrazing of kelp by sea urchins, affecting the entire kelp forest ecosystem (Estes *et al.*, 1989; Tegner and Dayton, 2000). Collapse of overexploited resources can produce severe economic hardship for workers in specialized industries that exploited the resources if no similar resources are available for exploitation, as occurred following the recent collapse of the Atlantic Cod *Gadus morhua* (Hutchings and Myers, 1994; Myers *et al.*, 1997*b*) and several other fisheries in Eastern Canada and the U.S.A.

Economic factors commonly contribute to deterministic causes of overexploitation. Unregulated competition, as in open-access fisheries in international waters, leads to a 'tragedy of the commons' in which competing harvesters each are driven to overexploit a resource before their competitors do (Hardin, 1968; Ludwig *et al.*, 1993). Even for exclusively or cooperatively managed resources Clark (1973, 1990) showed that when the economic discount rate exceeds the maximum growth rate of a population the optimal strategy from a narrow economic point of view is immediate liquidation of the population (harvesting to extinction) because money in the bank will grow faster than the population. Economic discount rates used by corporations and governments are typically about 5–10% per year or larger. This exceeds the maximum rate of population growth for many forest trees, and for typical vertebrates with adult body weights above 1 kg, which usually have $r_{max} < 0.1$ yr^{-1} (Charnov, 1993). Scarcity of a resource can increase its price, promoting increasing effort per unit catch as the total annual catch shrinks. Finally, in a recurring historical pattern, when exploitation of a particular resource generates large profits, harvesters become sufficiently wealthy and influential to induce governments to subsidize overexploitation of a diminishing resource (Ludwig *et al.*, 1993; James *et al.*, 1999). Natural resource economics and ecological economics provide solutions to these dilemmas by showing how to value resources for society as a whole rather than only for the resource extractors (Groombridge, 1992; Heywood, 1995; Tietenberg, 1998; Armsworth and Roughgarden, 2001).

Stochasticity and uncertainty in population dynamics also have contributed substantially to overexploitation. Harvesting strategies based on deterministic models, such as maximum sustained yield, are still commonly used but do not account for stochastic population dynamics, which can lead to overexploitation and resource collapse or extinction if harvesting continues when the population becomes small (Larkin, 1977). Uncertainty in population estimates can cause overexploitation because, when population sizes are overestimated, harvesting quotas will be set too high and the population may be reduced to dangerously low levels (Hilborn and Walters, 1992).

Despite these observations, relatively little effort has been devoted to understanding how stochastic population dynamics affects the risk of resource collapse or extinction. Much recent harvesting theory relies on deterministic or stochastic simulations of age- or stage-structured populations (Getz and Haight, 1989; Quinn and Deriso, 1999). Initial analytical models of harvesting fluctuating populations neglected age structure and assumed a stationary distribution of population size, considering extinction, if at all, using a simple condition for population persistence, i.e. that a stationary distribution exists (e.g. Beddington and May, 1977; Reed, 1978). Here we review recent analytical models of sustainable harvesting of fluctuating populations without age structure that incorporate the risk of population collapse or extinction. Using diffusion theory we compare three classical harvesting strategies and one new strategy in terms of their harvest statistics and the mean time to population collapse or extinction. We employ simple analytical models to derive general principles and then apply these principles to more realistic, age-structured models of particular species to derive by simulation the optimal harvesting strategies.

6.2 Harvesting fluctuating populations

6.2.1 Harvesting strategies and goals

We analyze a general form of stochastic population dynamics with harvesting,

$$\frac{dN}{dt} = \beta(N, t)N - y(N) \qquad 6.1$$

where $\beta(N, t)$ is the stochastic density-dependent rate of population growth without harvesting and $y(N)$ is the rate of harvest or yield from a population of size N (Lande *et al.*, 1995). The diffusion approximation to this model (Chapter 2.3) has infinitesimal mean $M(N) = M_0(N) - y(N)$ and infinitesimal variance $V(N)$, where $M_0(N)$ represents the expected dynamics in the absence of harvesting.

Three classical harvesting strategies are described in Figure 6.1 and Table 6.1. The constant harvest strategy removes individuals at a constant rate, a, regardless of population size. Proportional harvesting (or constant effort harvesting) removes individuals at a rate proportional to the population size, bN. Under threshold harvesting the population is exploited at the highest rate possible when it exceeds a certain size, c, with no harvest when the population is below c. Of these three classical harvesting strategies, constant harvesting and proportional harvesting are the simplest to perform because they require no information on population size, N, whereas threshold harvesting can be carried out only by using estimates of population size. Here we assume that population size is known exactly. The next section considers the influence of uncertainty in population estimates.

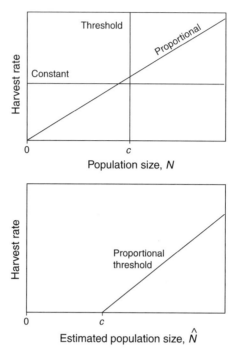

Fig. 6.1 Three classical harvesting strategies as functions of actual population size, N, and one new harvesting strategy as a function of estimated population size, $\hat{N}$. (From Lande *et al.*, 2001.)

Table 6.1 Classical exploitation strategies giving the harvest rate, y, as a function of population size, N

Strategy	Harvest rate[1]
Constant	$y(N) = a$
Proportional	$y(N) = bN$
Threshold	$y(N) = \begin{cases} 0 & \text{for } N < c \\ y_{max}(N) & \text{for } N \geq c \end{cases}$

[1] a, b, and c are constants that determine the harvest rate. The maximum harvest rate at population size N is $y_{max}(N)$.

It is essential to distinguish the goal of harvesting from the exploitation strategy. An extremely prudent goal is to maximize the expected cumulative yield over all time before eventual extinction of the population or reduction to a specified size. A more aggressive goal is to maximize the mean harvest rate that can be sustained over many years. A range of intermediate or more extreme goals also can be considered. For any given harvesting goal, each type of exploitation strategy can be optimized by choosing the best parameter value(s). Most actual harvesting is done with a goal similar to maximizing the mean sustainable harvest rate and a strategy resembling proportional (constant effort) harvesting. When thresholds are used, they typically are set too low, at the point of resource collapse or endangerment (e.g. Ludwig *et al.*, 1993; RSRP, 2001).

6.2.2 Yield statistics

For an initial population size N_0 the expected cumulative yield before extinction, $Y(N_0)$, is given by integrating the harvest rate function $y(N)$ with respect to the expected cumulative distribution of time spent at each population size before extinction, the Green function $G(N, N_0)$ (eq. 2.6a) evaluated using the infinitesimal moments of the diffusion approximation to equation 6.1,

$$Y(N_0) = \int_0^\infty y(N)G(N, N_0)\,dN. \qquad 6.2$$

The mean harvest rate along a typical population trajectory before extinction, $\bar{y}$, can be approximated by the expected cumulative yield divided by the mean time to extinction, $T(N_0)$, given by equation 2.6b,

$$\bar{y} \cong Y(N_0)/T(N_0). \qquad 6.3a$$

Similarly, the variance in harvest rate is approximately

$$\sigma_y^2 \cong \frac{1}{T(N_0)} \int_0^\infty [y(N)]^2 G(N, N_0)\,dN - \bar{y}^2. \qquad 6.3b$$

Exact formulas for the mean and variance of harvest rate would involve the integral of the harvest rate function and its square along particular population trajectories, each divided by the time to extinction for the population trajectory, averaged over the distribution of population trajectories. However, formulas 6.3 will be reasonably accurate if the cumulative yield is highly correlated with the duration of the sample paths, which will be true when harvesting allows the population to increase when rare so that the mean time to extinction is much longer than the expected duration of the final decline to extinction, $T(N_0) \gg T^*(N_0)$ (see Chapter 2.6, 2.7).

Figure 6.2 depicts typical population trajectories for the conditional and unconditional diffusion processes under the three harvesting strategies in Table 6.1 for a model with stochastic density-independent growth rate and deterministic density dependence of the logistic form, $dN/dt = \beta(t)N - \bar{\beta}N^2/K - y(N)$. Figure 6.3 shows the Green functions for the three harvesting models using the same parameter values as in Figure 6.2. As illustrated by these figures, in many situations $T(N_0) \gg T^*(N_0)$ so that the harvest rate statistics (eqs 6.3) have good accuracy.

The mean time to extinction, the expected cumulative yield before extinction and mean annual yield are compared in Figure 6.4 under the three harvesting strategies

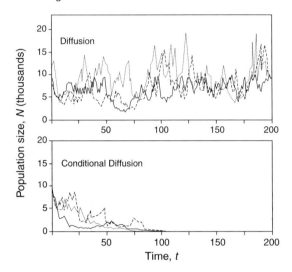

Fig. 6.2 Sample paths for the unconditional and conditional diffusion processes describing stochastic harvesting under three classical harvesting strategies with optimized parameters for the goal of maximizing the expected cumulative yield before extinction, with initial population size at carrying capacity, $N_0 = K$. Constant harvesting (dotted line), proportional harvesting (dashed line), threshold harvesting (solid line). Expected dynamics in the absence of harvesting are logistic with an Allee effect, $M(N) = \bar{\beta}(N - A)(1 - N/K) - y(N)$ with $\bar{\beta} = 0.1$ per year, $A = 20$, $K = 10,000$. Demographic and environmental stochasticity in density-independent growth rate takes the form $V(N) = \sigma_d^2 N + \sigma_e^2 N^2$ with $\sigma_d^2 = 1$ and $\sigma_e^2 = 0.04$. (From Lande *et al.*, 1995.)

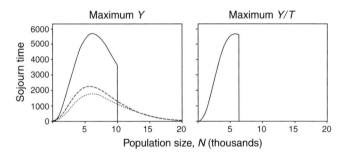

Fig. 6.3 Distributions of expected cumulative time spent at each population size before extinction (sojourn time or Green function, eq. 2.6a) corresponding to the unconditional sample paths under the three classical harvesting strategies as in Fig. 6.2. (From Lande *et al.*, 1995.)

for a range of harvest intensities. Threshold harvesting produces not only the highest mean sustainable harvest rate but also the largest expected cumulative yield before extinction. However, threshold harvesting also entails a high variance in harvest rate because of frequent periods of no harvest when the population size is below the

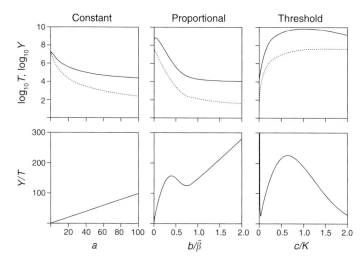

Fig. 6.4 Expected cumulative yield before extinction, Y (solid lines), mean time to extinction, T (dashed lines), and approximate mean annual yield, Y/T, for the three classical harvesting strategies as functions of the parameter values. Parameters for the proportional and threshold strategies are expressed relative to the mean intrinsic rate of increase, $\bar{\beta}$, and carrying capacity, K, respectively, with dynamics as in Figure 6.2. (Modified from Lande *et al.*, 1995.)

threshold. For the goal of maximizing the mean harvest rate under the threshold strategy, with an unlimited harvesting capacity and a large initial population size, the optimal threshold is $c = 0$, as shown in Figure 6.4, but this causes rapid extinction and is not sustainable. The optimal threshold that maximizes the mean sustainable harvest rate always occurs at a positive threshold.

Figure 6.4 suggests that under threshold harvesting with an unlimited harvesting capacity, the highest expected cumulative yield before extinction is achieved when the threshold is at the carrying capacity, $c = K$. The next section shows that this result is quite general and holds for a broad class of population dynamic models.

6.2.3 Optimal strategies for different goals

Consider the optimal strategies for the goals of maximizing Y or Y/T subject to the constraint $0 \leq y(N) \leq y_{max}(N)$. First express Y in an alternate form by substituting into equation 6.2 $y(N) = M_0(N) - M(N)$. The second term is then the expectation of $M(N)$ along sample paths, which can be evaluated using t^* to denote the time to extinction along a particular sample path, $N(t^*) = 0$, starting from $N(0) = N_0$,

$$\int_0^\infty M(N)G(N, N_0)\,dN = \mathrm{E}\left[\int_0^{t^*} \left(\frac{dN}{dt}\bigg|N\right)dt\right] = N(t^*) - N(0) = -N_0.$$

Thus equation 6.2 becomes

$$Y = N_0 + \int_0^\infty M_0(N)G(N, N_0)\, dN. \qquad\qquad 6.4$$

This is maximized when the Green function is as large as possible at population sizes where $M_0(N)$ is (highly) positive and as small as possible at population sizes where $M_0(N)$ is low or negative. The exploitation strategy that maximizes Y should therefore be such that the population spends as much time as possible at sizes where its expected growth rate without harvesting is high and as little time as possible at population sizes where its expected growth rate without harvesting is low or negative. This provides an intuitive explanation of why (with known population sizes) the optimal harvesting strategy for any goal always involves a threshold function in which the harvesting rate is either zero or as large as possible on either side of the threshold (Lande *et al.*, 1995). It was previously known for stochastic population models assuming a stationary distribution of population size that the optimal strategy for maximizing the mean harvest rate is a threshold harvesting function (Reed, 1978, 1979).

In most cases the optimal threshold harvesting strategy has a single threshold as in Table 6.1. This always applies when the expected dynamics in the absence of harvesting has a single stable equilibrium. A threshold strategy is also optimal or nearly optimal even when there is a second unstable equilibrium at low population density caused by an Allee effect. If the unstable equilibrium is sufficiently large then the optimal strategy may have a second threshold below the unstable equilibrium with harvesting at the maximum possible rate to remove the last remaining individuals from a population that almost certainly will decline to extinction (Lande *et al.*, 1995). However, this has little practical consequence for improving yields, and conservation efforts should instead be undertaken before this occurs. Of more interest is that under threshold harvesting an Allee effect far below the threshold can greatly decrease the expected cumulative yield and population longevity. For example, comparison of Figure 6.4 for threshold harvesting with Figure 6.5(A,B) for $\sigma_e = 0.2$ shows that an Allee effect decreases both the mean time to extinction, $T(K)$, and the expected cumulative yield before extinction, $Y(K)$, by more than an order of magnitude.

The optimal threshold, c, depends on the harvesting goal. For the conservative goal of maximizing the expected cumulative yield before extinction, Y, with unlimited harvesting capacity, $y_{\max}(N) = \infty$, the time spent by the population at sizes where $M_0(N)$ is positive is maximized and the time spent where $M_0(N)$ is negative is minimized when the threshold is at the stable equilibrium or carrying capacity, $c = K$. This remarkable result does not depend on the form of stochasticity or the expected dynamics, provided that the dynamics can be accurately approximated by a diffusion process (Lande *et al.*, 1995), as confirmed by Whittle and Horwood (1995). In contrast to traditional harvesting strategies, threshold harvesting with $c = K$ involves comparatively little disturbance to a harvested population and its ecosystem, harvesting only excess individuals above the carrying capacity that are likely to be eliminated by density-dependent regulation.

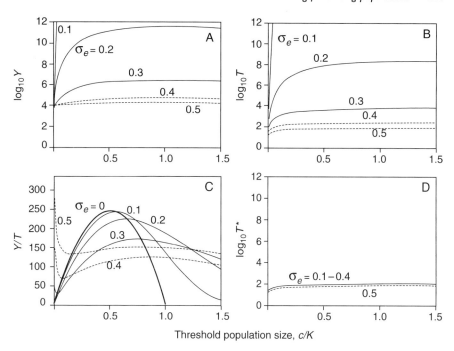

Fig. 6.5 (A) Expected cumulative yield before extinction, Y, (B) mean time to extinction, T, and (C) mean annual yield approximated by Y/T, as functions of the threshold population size, c, above which all excess individuals are harvested and below which no harvest occurs. Annual harvest in the deterministic model is indicated by the bold line in panel C. (D) Mean duration of the final decline from carrying capacity to extinction, T^* (Chapter 2.7). Expected dynamics without harvest are logistic, $M(N) = \bar{\beta}N(1 - N/K) - y(N)$ with $\bar{\beta} = 0.1$ per year, $K = 10{,}000$ and $N_0 = K$. Demographic and environmental stochasticity in density-independent growth rate takes the form $V(N) = \sigma_d^2 N + \sigma_e^2 N^2$ with $\sigma_d^2 = 1$ and values of the environmental standard deviation σ_e as shown. Dashed lines represent parameter values for which T is not much greater than T^*. (From Lande *et al.*, 1997.)

Analytical formulas for the optimal threshold under other harvesting goals have been derived (Lande *et al.*, 1995), but the simplest method of obtaining the optimal threshold for other goals is by numerical maximization of the quantity to be optimized. For the more usual goal of maximizing the mean annual yield, Figures 6.4 and 6.5 show that there are two optimal thresholds, one an edge optimum at $c = 0$ corresponding to immediate harvesting to extinction, and the other an internal optimum with $c > 0$ representing sustainable harvesting over several or many years. Under sustainable harvesting of a population with a positive long-run growth rate, the optimal threshold generally increases with larger environmental stochasticity, as illustrated in Figure 6.5(C). This result embodies the Precautionary Principle (O'Riordan and Cameron, 1995; O'Riordan *et al.*, 2001) as applied to sustainable harvesting. Exceptions to this general rule occur when the environmental stochasticity is sufficiently large that $T(K)$ is not much greater than $T^*(K)$ and the population does not persist very long. In this situation, corresponding to parameter values for the dashed lines in Figures 6.5 and 6.6, larger environmental stochasticity reduces the optimal threshold (Figs 6.5B,C, 6.6B), suggesting that more intense harvesting is necessary to maximize the mean annual yield before the population becomes quickly extinct. Exceptions also occur under extreme forms of density dependence, as in the theta-logistic model with very large values of θ (about 100) (eq. 5.1, Fig. 5.1B), approaching the model of density-independent growth to a ceiling (Sæther *et al.*, 1996*b*).

With a limited harvesting capacity, under any harvesting goal, decreasing the maximum harvesting capacity (e.g. by limiting the size of a fishing fleet) allows the population to exceed the threshold before it is reduced by harvesting at the maximum rate. This reduces the optimal threshold in comparison to the case of unlimited harvesting capacity (Lande *et al.*, 1995, 1997).

6.2.4 Uncertainty in population size

Sizes of harvested populations usually are estimated with substantial uncertainty. Coefficients of variation due to sampling often are in the range of 10–50% as shown in Table 6.2. Engen *et al.* (1997) analyzed the influence of uncertainty in population estimates on threshold harvesting strategies, assuming an unlimited harvesting capacity and that an estimated population size, $\hat{N}$, is available once each year, such that for a given actual population size the estimates are approximately normally distributed. Pulsed annual harvesting is assumed to occur at the times when population estimates are made, with harvesting strategy based on the estimated population size, $y(\hat{N})$.

With uncertainty in estimated population size, $\hat{N}$, information about the actual population size, N, is contained in the quasi-stationary joint distribution of actual and estimated population sizes. Approximating $E[N|\hat{N}]$ as a linear regression with slope q

Table 6.2 Coefficients of variation (CV) of estimates of population size for species of fish, birds, and large mammals

Group	Species	CV	Reference
Fish	Haddock *Melanogrammus aeglefinus*	0.19–0.26	Lowe and Thompson (1993)
	Various ground fish	0.20–0.50	Overholtz *et al.* (1993)
Birds	Various song birds	0.11–0.27	Buckland *et al.* (1993)
Mammals	Moose	0.31–0.56	Crête *et al.* (1986)
	Feral Horse *Equus caballus*	0.04–0.25	Eberhardt (1982)
	Spotted Dolphin *Stenella caeruleoalbus*	0.09–0.25	Buckland *et al.* (1993)
	Minke Whale *Balaenoptera acutorostrata*	0.15	Schweder and Holden (1994)
	Fin Whale *Balaenoptera physalus*	0.16	Buckland *et al.* (1993)

between 0 and 1 suggests a *proportional threshold harvesting* strategy. This involves harvesting a proportion q of the estimated population above the threshold each year, with no harvest when the estimated population is below the threshold (Fig. 6.1),

$$y(\hat{N}) = \begin{cases} 0 & \text{for } \hat{N} < c \\ (\hat{N} - c)q & \text{for } \hat{N} \geq c. \end{cases} \qquad 6.5$$

With uncertainty in population estimates, setting $q < 1$ reduces the risk of over-harvesting in years when the population size is overestimated.

Clark and Kirkwood (1986) employed dynamic programming to include uncertainty in $\hat{N}$ in the stochastic model of Reed (1979). In numerical examples they showed that the optimal harvest rate increased approximately linearly above a threshold, as in equation 6.5.

Engen *et al.* (1997) analyzed a model with a stochastic density-independent growth rate and a general deterministic form of density dependence,

$$\frac{dN}{dt} = \beta(t)N - f(N) - y(\hat{N}) \qquad 6.6$$

The diffusion approximation to this model has infinitesimal mean and variance

$$M(N) = \bar{\beta}N - f(N) - E[y(\hat{N})|N] \quad \text{and}$$

$$V(N) = \sigma_e^2 N^2 + \sigma_d^2 N + \text{Var}[y(\hat{N})|N] \qquad 6.7a$$

where σ_d^2 and σ_e^2 are, respectively, the demographic and environmental variance in the density-independent growth rate. Now assume that the distribution of $\hat{N}$ for a given value of N, denoted as $F(\hat{N}|N)$, is approximately normal with mean N and variance $U^2 N$, where U is a constant depending on the method and intensity of sampling and the spatial clumping of the population (Seber, 1982; Buckland *et al.*, 1993). Then under the proportional threshold harvesting strategy (eq. 6.5) we have

$$E[y(\hat{N})|N] = q \int_c^\infty (\hat{N} - c) F(\hat{N}|N) \, d\hat{N} \qquad 6.7b$$

and

$$\text{Var}[y(\hat{N})|N] = q^2 \int_c^\infty (\hat{N} - c)^2 F(\hat{N}|N) \, d\hat{N} - \left\{ E[y(\hat{N})|N] \right\}^2. \qquad 6.7c$$

The integrals in these expressions produce explicit formulas in terms of ordinates the standard normal distribution and its integral (Engen *et al.*, 1997).

Figure 6.6 illustrates the joint optimization of the threshold c and the proportion, q, as a function of the amount of uncertainty measured as the coefficient of variation in estimated population size when the actual population size is at carrying capacity, $U_K = U/\sqrt{K}$. For known population sizes, $U = 0$, with the goal of maximizing Y the optimal threshold is below K and decreases with increasing environmental

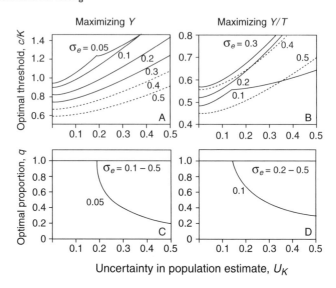

Fig. 6.6 Optimal harvesting threshold relative to carrying capacity, c/K, and optimal proportion, q, of the estimated population above c harvested annually (A,C) to maximize Y or (B,D) to maximize Y/T as functions of the uncertainty in estimated population size, U_K, the coefficient of variation of estimate when actual $N = K$. Other parameters as in Figure 6.5 with environmental standard deviation $\sigma_e = 0.2$. (From Lande *et al.*, 1997.)

stochasticity. This result for pulsed annual harvesting contrasts with the previous result for continuous-time harvesting with known population size, under which the optimal threshold always is $c = K$. This occurs because with pulsed annual harvesting the population may exceed K before the next harvest, reducing Y due to density-dependent losses, thus lowering the optimal threshold. This effect of pulsed annual harvesting also applies with uncertainty. For the goals of maximizing either Y or Y/T the optimal threshold increases with increasing uncertainty. With high uncertainty and low environmental stochasticity, the optimal q may be substantially less than 1.

The above analysis of uncertainty suggests a method for reducing the high variance in harvest rate associated with classical threshold harvesting. Figure 6.6 indicates that when the optimal q is less than 1, this can substantially lower the optimal threshold, reducing the periods with no harvest. Provided that $T(K) \gg T^*(K)$, even when the optimal $q = 1$ the mean annual yield, $\bar{y} \cong Y/T$, and the mean time to extinction starting from carrying capacity, $T(K)$, are not greatly reduced by changes in q. Thus a good method for reducing variance in harvest rate is to choose a low value of q, about 0.1, while optimizing the threshold, c. For a given choice of $q \gg \sigma_e^2$, with known population sizes, $U = 0$, the optimal threshold for maximizing Y under pulsed annual harvesting, is approximately independent of the form of density dependence,

$$c \cong \left[1 - \sigma_e \sqrt{\frac{\pi}{4q}} \right] K. \qquad 6.8$$

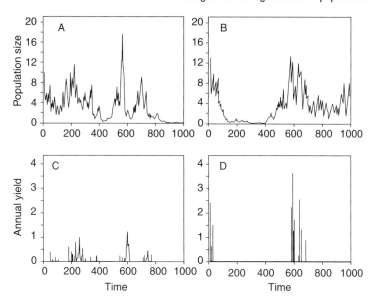

Fig. 6.7 Simulation of population size and pulsed annual harvest for two values of q under threshold harvesting with c chosen to maximize Y, the expected cumulative yield before extinction. Parameters are (A,C) $q = 0.1, c/K = 0.59$ and (B,D) $q = 1, c/K = 1.04$. Uncertainty in estimated population size is $U_K = 0.2$, the coefficient of variation of the estimate when $N = K$. Expected dynamics in the absence of harvest are logistic with $\bar{\beta} = 0.01$ and $K = 10,000$. Density-independent growth rate has demographic variance $\sigma_d^2 = 1$ and environmental variance $\sigma_e^2 = 0.025$. (From Engen *et al.*, 1997.)

Figure 6.7 compares sample paths of pulsed annual harvests in a population model for which maximization of the mean annual yield under joint optimization of q and c produces $q = 0$, versus preassigning $q = 0.1$ and optimizing only c.

Aanes *et al.* (2002) accounted for uncertainty in estimated population size and in population parameters to analyzed sustainable harvesting strategies for the Willow Ptarmigan (*Lagopus lagopus*), an important game species in Fennoscandia, that begins reproduction at age 1 year, justifying their use of population models neglecting age structure. The concepts of threshold and proportional threshold harvesting, derived from simple analytical models without age structure, are extended to age-structured populations in the following examples, using simulation to explore the optimal strategies of sustainable harvesting (Lande *et al.*, 2001).

6.3 Harvesting stochastic age-structured populations

6.3.1 Scandinavian Brown Bear

The Brown Bear was persecuted in Scandinavia for several centuries because it preyed on sheep and semi-domestic reindeer. The bear was essentially exterminated in Norway, and by 1930 about 100 individuals remained in Sweden (Swenson *et al.*, 1995). After conservation measures by the Swedish government, the population increased, resulting in immigration of individuals into Norway. There are now about 1,000 brown bears in Sweden and Norway (Swenson *et al.*, 1998). Following removal of large predators, Norwegian livestock breeders now practice less intensive husbandry, leaving sheep unattended. Immigrating Brown Bears can therefore cause serious problems, in some areas killing up to 10% of the ewes, creating strong political and public pressure for killing individual problem bears. In such areas, the allowable size of bear populations will be severely restricted, which may conflict with an obligation to maintain viable populations under the Convention on Biological Diversity (1992).

Control of Brown Bear populations should employ a strategy that minimizes the risk of population collapse or extinction. Previous analyses of Scandinavian Brown Bear demography based on data from radio-collared individuals demonstrated large uncertainty in several demographic parameters, resulting in rather uncertain predictions of future population sizes (Sæther *et al.*, 1998*b*). Precise estimates of population size also are difficult to obtain for this species. Both types of uncertainties must be considered when proposing management strategies for Brown Bear. A proportional threshold harvesting strategy has been suggested to account for such uncertainties in the management of the Norwegian Brown Bear (Tufto *et al.*, 1999). Here we employ a proportional threshold harvesting strategy and calculate the minimum harvesting threshold, c, to maintain a viable population according to criteria of the World Conservation Union (Table 5.1; IUCN, 2001).

We analyzed the population dynamics of the Norwegian Brown Bear using a density-independent model without age structure (eq. 5.2b). Estimates of demographic parameters and their uncertainties were derived from the density-independent age-structured model of Sæther *et al.* (1998*b*) by parametric bootstrapping. With a preassigned value of q in the proportional threshold harvesting strategy, for each bootstrap replicate of the population parameters, a corresponding bootstrap replicate of the threshold, $\tilde{c}$, was obtained, satisfying the criterion that the probability of extinction before 100 years is 0.10. The upper 95% quantile of the distribution of $\tilde{c}$ was chosen to be precautionary as the probability that the correct threshold c is larger than this value is approximately only 5%. This analysis is illustrated in Figure 6.8, showing two important points. First, with increasing uncertainty in estimated population size, a larger threshold c is required to avoid extinction with the prescribed probability. Accurate population estimation is therefore a prerequisite for removing individuals from such a small population. Second, the smallest population size at which harvest is allowable depends on q, the proportion of the excess estimated population above the threshold that is removed. A small q permits harvesting at a lower population size. Alternative management strategies that differ in q and c may produce the same probability of extinction with approximately 95% confidence. For example, the goal of minimizing the expected equilibrium population size would be achieved by setting a low q, but when population estimates have large uncertainty, the threshold c must be substantially increased to prevent extinction. Then the

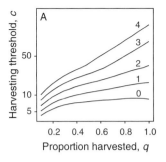
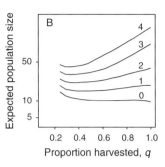

Fig. 6.8 The upper 95% quantile of the harvesting threshold, c, and the expected population size at the quasi-stationary distribution produced by a balance between density-independent growth (eq. 5.2a) and proportional threshold harvesting, as functions of the preassigned q, the proportion harvested above the threshold, for Norwegian Brown Bear. Numbers above each line indicate different levels of uncertainty in population estimates, U, where $U/\sqrt{N}$ is the coefficient of variation of estimated population size due to sampling errors. Each line gives a 10% extinction risk in 100 years with approximately 95% confidence. Population parameters are estimated from the Brown Bear in southern Sweden, $\bar{\beta} = 0.15$, $\sigma_d^2 = 0.155$, and $\sigma_e^2 = 0.003$ (Sæther *et al.*, 1998b). (From Lande *et al.*, 2001.)

smallest expected population size is produced by an intermediate q equal to 0.35. This illustrates the importance of including uncertainty in population estimates in the formulation of harvesting strategies for small populations. However, such small populations of Norwegian Brown Bear (Fig. 6.8) would not be either genetically (Lande, 1995) or demographically viable in the long run, were it not for occasional immigration from the much larger population in Sweden.

6.3.2 Fennoscandian Moose

The most important game animal in Fennoscandia is the Moose. After World War II the population and the harvest increased continuously until 1982 when more than 170,000 individuals were harvested in Sweden alone. The annual harvest in Finland, Sweden, and Norway combined in 2001 was nearly 210,000 individuals. Although moose meat is an important commodity in the region, a large moose population also exerts costs to forestry operations by browsing on commercially important tree species and by accidental collisions with cars and trains.

Moose population dynamics are strongly influenced by environmental stochasticity (Solberg *et al.*, 1999). Body weight of the youngest females is affected by variation in climate, especially summer precipitation (Sæther, 1985; Solberg and Sæther, 1994) and winter snow depth (Sæther *et al.*, 1996a). Body weight then influences age at maturity because larger females have a higher chance of conceiving as yearlings (Sæther and Heim, 1993; Post and Stenseth, 1999). Similar weight-dependent effects of climate occur in other ungulates (Sæther, 1997). Large annual variation also is found in neonatal survival of calves in northern populations (Sæther *et al.*, 1996*a*; Stubsjøen *et al.*, 2000). It is therefore essential to incorporate environmental stochasticity into management plans for moose populations.

We analyzed an age-structured model with n_i females and m_i males in age class i, where $i = 0, 1, \ldots, 20$. Change in numbers occurs sequentially in three stages: reproduction occurs in spring, the population is harvested during autumn, and death by natural causes happens during winter. At the end of winter, individuals move from one age class to the next. Climatic factors and density regulation affect winter survival of calves as well female reproduction at age 2 (see Sæther *et al.*, 2001).

Hunting quotas of Scandinavian Moose are set as age- and sex-specific harvests in three categories: calves and adult (≥ 1 year old) males and females. Assuming known population parameters, we compared proportional harvesting and threshold harvesting strategies generalized to include uncertain age and sex structure. The proportional harvest strategy specifies the proportion of the estimated size to be harvested for each of the three categories of harvest. The threshold harvesting strategy specifies a threshold for each of the three categories, allowing removal of all excess individuals above each threshold. The optimal parameters for each strategy that produce the maximum mean annual yield are determined by repeatedly simulating the process over a large number of years. Using a fixed random number seed to initiate each simulation the mean annual yield is simply a function of the parameters. Optimization can then be performed using standard numerical procedures for maximizing functions of several variables.

Numerical results from this complex density-dependent age-structured model confirm the main analytical results from simple diffusion models without age structure. The threshold strategy generally produces a larger mean annual yield than the proportional harvesting strategy, although because of the high expected population growth rate of Scandinavian Moose, there is not much difference in the maximum mean annual yield under the two strategies, as shown in Figure 6.9. However, the corresponding variance in annual harvest is substantially larger under threshold harvesting than under proportional harvesting.

Calf survival is one of the most variable demographic traits in ungulate population dynamics (Sæther, 1997; Gaillard *et al.*, 1998, 2000) and is often highly correlated with annual variation in population growth rate. To examine how a reduction in the mean population growth rate would affect the harvest strategy, we decreased the natural calf survival in the model from 0.8 to 0.4. Such low survival rates may occur in areas where large carnivores prey heavily on moose calves (Larsen *et al.*, 1989; Ballard *et al.*, 1991). Low calf survival resulted in a larger difference in the mean annual yield between threshold and proportional harvesting strategies.

6.4 Reducing variance of annual yield in threshold harvesting

With known population sizes, for any harvesting goal a threshold harvesting strategy always produces a higher yield with a lower risk of resource collapse or extinction

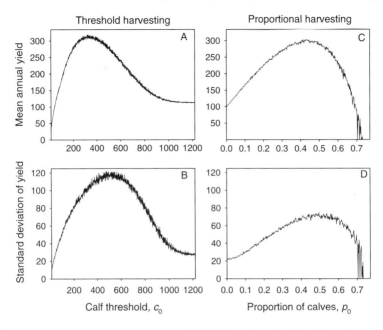

Fig. 6.9   Threshold harvesting: mean annual yield (A) and standard deviation of annual yield (B) as functions of the harvesting threshold for moose calves c_0, for fixed optimal values of the thresholds for adult (≥ 1.5 year old) males $c_m = 296$ and adult females $c_f = 1425$. Proportional harvesting: mean annual yield (C) and standard deviation of annual yield (D) as functions of the proportion of moose calves harvested annually, p_0, with proportions harvested annually for adult males $p_m = 0.182$ and adult females $p_f = 0.005$ fixed at their optimal values. The threshold harvesting model and the proportional harvesting model are based on a stochastic density-dependent age-structured model of moose with the same population parameters (Sæther *et al.*, 2001). (Modified from Lande *et al.*, 2001.)

than any other strategy. The mean time to extinction may be orders of magnitude larger under threshold harvesting than under constant or proportional harvesting (Fig. 6.4), because harvesting rates at small population sizes strongly influence the mean time to extinction. Threshold harvesting allows a population to recover at its maximum natural rate when it is below the threshold. Despite these advantages threshold harvesting is rarely employed in resource management because it entails a high variance in annual yield, with frequent years of no harvest when the population is below the threshold. The historical failure of resource management agencies to exercise sufficient caution to prevent frequent overexploitation and collapse of resources is caused by a prevailing philosophy of resource extraction over resource conservation, which, even in the presence of high stochasticity and uncertainty, favors continued harvests in the short term over resource conservation and sustainability in the long term (Ludwig *et al.*, 1993; RSRP, 2001).

Two features of the threshold harvesting and proportional threshold harvesting strategies help to reduce the variance in annual yield. First, under classical threshold harvesting, limiting the harvesting capacity reduces the optimal threshold. Second, proportional threshold harvesting reduces both the threshold and the harvest rate. These features allow the population to fluctuate, and harvesting to occur, over a wider range of population sizes. This reduces the variance in annual yield and the frequency of years with no harvest. It also facilitates the estimation of population parameters necessary to develop optimal harvesting strategies.

Financial remedies that can ameliorate a high variance in annual harvest include banking a proportion of profits from years of high harvest and insurance against low harvest. Profit-sharing among harvesters specializing on different resources that fluctuate asynchronously also would help to reduce variance in annual income.

Support is increasing for the establishment of large reserve areas where harvesting is prohibited, while allowing continuing harvests outside the reserves, especially for marine systems (Schrope, 2001). Marine reserves benefit many harvested target species, and large-bodied species subject to bycatch mortality (Mosqueira *et al.*, 2000), but may not prevent overexploitation and population collapse for species with high dispersal rates unless a substantial fraction of the relevant habitat is reserved from harvest (Clark, 1996; Botsford *et al.*, 2001). Large reserves help to improve population persistence and increase reliability of harvests in comparison with strategies of continued harvesting without reserves (Mangel, 2000). However, reserves alone, with continued intense harvesting outside reserves, are not as precautionary as threshold harvesting or proportional threshold harvesting involving complete cessation of harvests in years of low abundance. Nevertheless, even relatively small reserves can substantially improve sustainablility of harvests if they are sited in especially favorable habitat that serves as population sources, or in localities essential for life cycle continuity such as breeding areas (Garcia-Charton and Perez-Razafa, 1999). A combination of reserves and threshold harvesting strategies could do much to promote viable populations and sustainable harvests for exploited species.

6.5 Summary

Many harvested species are overexploited because of increasing demand from growing human populations, economic factors, and inadequate regulation that often fails to account for stochasticity and uncertainty in resource dynamics. The goal of harvesting should be distinguished from the harvesting strategy. With known population sizes, the optimal strategy for any goal generally involves threshold harvesting in which the population is harvested at the maximum possible rate when it is above the threshold, with no harvesting below the threshold. This generally produces a higher average or cumulative harvest with a lower risk of population collapse or extinction than alternative strategies involving continual harvesting, because the population is allowed to recover at its maximum natural rate of increase below the threshold. For the extremely conservative goal of maximizing the expected cumulative harvest before eventual extinction, under any form of stochasticity that can be accurately modeled as a diffusion process, the optimal strategy is immediate harvesting of all individuals above the carrying capacity with no harvest below the carrying capacity. For the more usual goal of maximizing the mean annual yield sustainable over many years, the optimal threshold usually exceeds that which would produce the maximum sustainable yield in a deterministic model. In populations that can persist for long times, larger environmental stochasticity usually increases the optimal harvesting threshold. Population sizes of exploited species typically are estimated with substantial uncertainty. With uncertain population sizes, a superior strategy often involves proportional threshold harvesting in which only a proportion of the population above a threshold is harvested each year, with no harvesting below the threshold. These principles can be applied to age-structured populations using computer simulations to derive optimal age- and sex-specific harvesting thresholds for any given harvesting goal.

In comparison with typical strategies involving continual harvesting, the major drawback of threshold harvesting is a relatively high variance in annual yield with frequent years of no harvest. This can be alleviated by limiting the harvesting capacity and by adopting proportional threshold harvesting with a suitable choice of the proportion harvested above the threshold. Both of these remedies reduce the optimal threshold, allowing limited harvesting to occur over a wider range of population sizes. Sustainability of renewable resources also can be improved by insurance and profit-sharing schemes, and by combining reserves with threshold harvesting.

7

Species diversity

Patterns of species diversity in space and time, and their causes, are central topics in ecology (MacArthur, 1972; Huston, 1994; Rosenzweig, 1995). Species diversity features prominently in prioritizing areas for conservation action, and nongovernmental conservation organizations commonly use 'biodiversity hotspots' to guide their resource allocation and planning (Conservation International, 2002a; Redford et al., 2002). However, assessing which country or area has the greatest species diversity is not as straightforward as it might seem, and many reports of biodiversity hotspots fail to account for differences in sample area or sampling effort. Only by developing formal statistical models of species abundance distributions and different measures of species diversity can we have the tools necessary for converting rapid ecological assessments and time-limited sampling into reliable evaluations of biodiversity.

This chapter introduces the species abundance distribution in a community or random sample. We describe statistical procedures that account for sampling variance in the estimated abundance of each species to fit species abundance distributions with Gamma or lognormal forms (Fisher et al., 1943; Preston, 1948). Statistical properties are described for the three most commonly used nonparametric scalar measures of species diversity: species richness, Shannon information, and Simpson diversity. For each of these diversity measures we analyze the statistical accuracy of estimates of the actual diversity in a community that can be made from random samples. We explain additive partitioning of species diversity into components attributable to subdivisions of the community in space and time. We conclude this chapter with examples of partitioning diversity into spatial and temporal components in highly diverse insect communities. The final chapter extends the single species models from previous chapters to derive the species abundance distribution and the stochastic dynamics of a community of interacting species in a single trophic guild; we apply this theory to data on a tropical butterfly community to analyze the causes of spatial and temporal variation in abundance among species.

7.1 Species abundance distributions

Species lists often are compiled to measure and compare species diversity among areas at a particular time. More detailed, but still static, descriptions of community structure include the relative abundance of species. A species abundance distribution plots the number of species $\Psi(x)$ as a function of their observed abundance, x, in a sample from a community. Preston (1948, 1962, 1980) emphasized that in large samples from diverse natural communities, species abundance distributions often are truncated on the left below a 'veil line' by omission of species too rare to be sampled (see also Williams, 1964). Following Preston, most authors use observed abundance intervals of log base 2 for species abundance distributions, with edges at 1/2 and $2^j + 1/2$ for $j = 0, 1, 2, \ldots$, containing $1, 2, 3-4, 5-8$, $9-16, \ldots$ individuals sampled. For graphical presentations only, we follow Williams (1964) who used intervals of log base 3 with edges at $3^j/2$, containing $1, 2-4, 5-13, 14-40, \ldots$ individuals sampled per species. However, in estimating parameters of species abundance distributions, particularly when these are lognormal, lumping the data into intervals is not necessary and we use natural logarithms (base e). We emphasize that the choice of logarithmic base influences the shape of species abundance distributions, particularly the existence and position of an internal peak. A larger logarithmic base progressively compresses the abscissa at large abundances, correspondingly raising the distribution and shifting its peak to the right.

Figure 7.1 illustrates a species abundance distribution for a large sample of fruit-feeding nymphalid butterflies from Amazonian Ecuador. The substantial number of species represented by one or a few individuals in this large sample suggests that other rare species exist in the community but are not present in the sample. A rank abundance plot giving the (logarithmic) relative abundance or frequency of each species as a function of its abundance rank is another way of portraying the same information as in the species abundance distribution (Fig. 7.1). The rank of a species' abundance in a community equals the number of species with larger or equivalent abundance, so the rank abundance plot is just the species abundance graphed against the upper cumulative species abundance distribution (May, 1975),

$$Rank = \sum_{x=N}^{\infty} \Psi(x). \qquad 7.1$$

A variety of parametric forms of species abundance distributions have been suggested, based on simple statistical principles or niche partitioning models (reviewed by May, 1975; Pielou, 1975; Engen, 1978; Magurran, 1988). Motomura (1932) proposed a model of sequential niche partitioning where the first species pre-empts a certain fraction of the resource space, and each subsequent species pre-empts the same fraction of the remaining resources. Assuming that the population size of each species is proportional to the amount of resource in its ecological niche, this obviously produces a geometric series of species abundances, which has been used to describe communities with low numbers of species such as subalpine plants (Whittaker, 1970). MacArthur (1960) derived the 'broken stick' model of species abundances in which a one-dimensional resource interval is divided among species by random placement of points on the interval. Motomura's niche pre-emption model and MacArthur's broken stick model are special cases of the general Gamma model of species abundance distribution (Engen, 1975, 1978). Similar mechanisms also can generate a lognormal distribution of species abundances if the niche space is sequentially broken in two

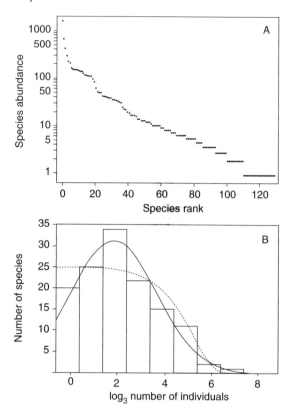

Fig. 7.1 Rank abundance plot (A) and species abundance distribution using log base 3 intervals (B) for a sample of 6,690 individual fruit-feeding nymphalid butterflies collected in 1 year of monthly sampling at Jatun Sacha, Ecuador. In panel B the solid curve is the lognormal distribution and the dashed curve is Fisher's logseries distribution fitted using the methods of Pielou (1975). (Modified from DeVries *et al.*, 1997.)

(with a fixed or random fraction) and the next piece to be broken is chosen randomly (Bulmer, 1974; Pielou, 1975; Sugihara, 1980). Pielou (1975) emphasized that these ecological mechanisms of partitioning species into static nonoverlapping niches are so simplistic as to be little better than purely statistical descriptions of species abundance distributions. The two most commonly used parametric forms of species abundance distributions are the Gamma and the lognormal, and sampling distributions derived from them, which we briefly describe below.

In general, in a community of S species their relative abundances p_1, p_2, $p_3, \ldots, p_S$ can be considered as random variates from the hypothetical species abundance distribution $\Phi(p)$ normalized so that $S \int_0^\infty p\Phi(p)\,dp = 1$. This condition only entails that the mean relative abundance of species in the community is $\bar{p} = 1/S$, and does not guarantee that relative abundances for S species sampled from this distribution will sum to 1, as must the observed frequencies in a sample. Defining the sampling intensity I as the expected total number of individuals in a random sample from the community, this can be estimated by n, the total number of individuals actually sampled, $\hat{I} = n$. For a particular species with relative abundance p the expected number of individuals in a random sample is Ip and the probability that this species is represented by x individuals is the Poisson formula $e^{-Ip}(Ip)^x/x!$. The expected species abundance distribution observed in a random sample of individuals from the community is the probability that any species is represented by x individuals in the sample,

$$\Psi(x) = S \int_0^\infty e^{-Ip} \frac{(Ip)^x}{x!} \Phi(p)\, dp. \qquad 7.2$$

7.1.1 Gamma and negative binomial

Fisher *et al.* (1943) proposed that the relative abundance of species in a community can be described by compounding a Gamma distribution of actual species abundances with a Poisson sampling distribution of individuals from each species, assuming the community contains a very large (effectively infinite) number of individuals and individuals are randomly sampled from the community. A Gamma distribution of relative abundances of species has shape and scale parameters k and $\alpha = Sk$, so that $\Phi(p) = [\alpha^k / \Gamma(k)]p^{k-1}e^{-\alpha p}$, where $\Gamma(k)$ is the Gamma function with the property $\Gamma(k+1) = k\Gamma(k)$ and $\Gamma(1) = 1$ so that for integers x the Gamma function is a simple factorial, $\Gamma(x) = (x-1)!$. This produces an expected distribution of observed species abundances that is a negative binomial distribution,

$$\Psi(x) = \alpha \frac{\Gamma(k+x)}{\Gamma(k+1)} \frac{(1-\omega)^k \omega^x}{x!}$$

where $\omega = I/(\alpha + I)$. The parameters k and α can be estimated from an observed species abundance distribution using maximum likelihood.

Fisher noticed that in samples from communities containing a large number of species, particularly tropical and temperate insect communities, k is often statistically indistinguishable from zero. Fisher's logseries model is the limiting version of the negative binomial distribution when $k \to 0$ and $S \to \infty$ such that $\alpha = Sk$ approaches a positive constant,

$$\Psi(x) = \alpha \frac{\omega^x}{x}.$$

This distribution is often used to describe highly diverse communities with an indeterminably large (effectively infinite) number of species, and with intense sampling of such communities ω typically is slightly less than unity. In a rank abundance plot of log abundance versus species rank in the community, Fisher's logseries model always produces an approximately linear relationship, within the constraints dictated by integer numbers of individuals sampled in each species, as can be inferred from the above formula with equation 7.1, and shown more rigorously by Engen (1975).

When presented as a species abundance distribution Fisher's logseries model is not expected to display a modal value, but instead tends to be nearly flat for species of low abundance as shown in Figure 7.1.

7.1.2 Lognormal and Poisson-lognormal

In natural communities with high species richness the species abundance distribution often is approximately lognormal (Preston, 1948, 1962, 1980; Williams 1964). When the distribution of relative abundance of species in a community is lognormal, with $\ln p_i$ having mean μ and variance σ^2, for S species in the community the normalization of $\Phi(p)$ above equation 7.2 implies $\exp\{\mu + \sigma^2/2\} = 1/S$, so that $\mu = -\ln S - \sigma^2/2$. Random sampling of individuals in the community produces, in equation 7.2, a Poisson-lognormal distribution of observed species abundances, which was analyzed by Grundy (1951) and Bulmer (1974) who suggested statistical methods for estimating the parameters S and σ^2. Pielou (1975) developed a simplified procedure for parameter estimation, but did not account for sampling effects.

The influence of sampling intensity on the expected shape of observed species abundance distributions is illustrated in Figure 7.2 for hypothetical communities in which the actual species abundances are lognormal or Gamma distributed. This illustrates the approximate nature of Preston's 'veil line' as the distributions for different sampling intensities are not simply truncated on the left, but are also slightly altered in shape above the veil line. It also shows that communities with a Gamma distribution of species abundance have proportionately more rare species than communities with a lognormal distribution of species abundance.

7.1.3 Nonrandom sampling from overdispersion

We have so far assumed that individuals are randomly sampled from a community. Several mechanisms can cause conspecific individuals to be nonrandomly distributed with respect to spatial and temporal sampling, which will bias the species abundance distribution. Deterministic and stochastic variation in the environment are the most important causes of nonrandom distributions of individuals within species. Other mechanisms that typically operate on small spatial and temporal scales include social aggregation and demographic stochasticity in small local populations. Territoriality in animals and allelopathy in plants can produce spatial distributions of conspecific individuals that are more uniform than expected by chance (underdispersion), but the net effects of various mechanisms usually result in conspecific clumping (or overdispersion) manifested as spatial and temporal variation in local abundance. In Chapter 8 we show how models of community dynamics in space and time can be used to partition the total variance in species abundances into different causes. We now analyze a simple model to investigate how intraspecific overdispersion in space influences parameters of local species abundance distributions (prior to individual sampling).

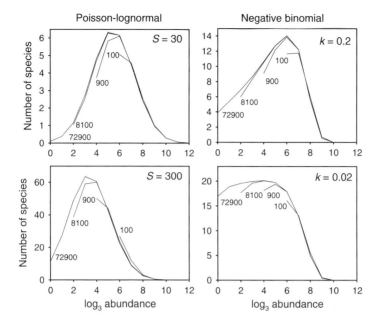

Fig. 7.2 Expected species abundance distributions for theoretical communities with lognormal or Gamma distributions of species abundance under different sampling intensities indicated by expected sample sizes. Individuals are assumed to be randomly sampled from each community. For expected sample sizes successively smaller by a factor of 3^{-2} the distributions are shifted to the right by 2 units on the $\log_3$ scale to facilitate comparison of their shapes.

For a given species with relative abundance p in a community, its relative abundance at a particular locality is denoted as $\tilde{p}$, writing the distribution among localities as $h(\tilde{p}|p)$ with mean p and variance $c^2 p^2$. The coefficient of variation of relative local abundance, c, measures the overdispersion. The expected species abundance distribution at a locality is $\int_0^\infty h(\tilde{p}|p)\Phi(p)\,dp$ from which random sampling of individuals will produce the observed local species abundance distribution $\Psi(x)$ (cf. eq. 7.2). Consider for example a lognormal species abundance distribution in a community in some region, and assume that intraspecific overdispersion occurs independently among species, with the same magnitude in each species. Write the local relative abundance of a given species as proportional to its regional relative abundance, $\tilde{p} = up$, and assume that among localities the proportionality factor u is lognormally distributed with mean 1 and variance c^2. Then at a given locality $\ln \tilde{p} = \ln u + \ln p$ is normally distributed among species with mean and variance (cf. eq. 1.5a)

$$\mu_{\ln \tilde{p}} = \mu_{\ln p} - \tfrac{1}{2}\ln(1 + c^2) \quad \text{and} \quad \sigma^2_{\ln \tilde{p}} = \sigma^2_{\ln p} + \ln(1 + c^2).$$

Thus intraspecific overdispersion is expected to decrease the mean log abundance and increase the variance of log abundance among species at a locality. Because $\sigma^2_{\ln p}$ is often in the range of about 2–4 (Fig. 7.1; Preston, 1948, 1962, 1980; Williams 1964), moderate or high intraspecific overdispersion ($c > 0.3$) can substantially broaden local species abundance distributions.

7.2 Measures and statistics of diversity

Indices of species diversity have a long history, and appear frequently in descriptions of spatial patterns in biogeography and community ecology (MacArthur, 1972; Huston, 1994; Rosenzweig, 1995). Diversity indices are routinely employed as measures of community structure in analyzing anthropogenic impacts on the environment, and they are now often used by conservation organizations in evaluations of biodiversity (Conservation International, 2002a; Redford *et al.*, 2002). The following sections compare the statistical properties of commonly used measures of species diversity, assuming for simplicity that individuals are randomly sampled from a community. We emphasize that in a diverse community observed species richness has a large bias caused by its strong dependence on sample size. Rarefaction is explained as a statistical method of accounting for differences in sample size in comparisons of species richness. Finally, we show how to partition a measure of species diversity into additive components attributable to different factors, and how to assess the statistical significance of components of diversity.

Any measure of species diversity should satisfy some basic properties. These include *non-negativity* so that a negative diversity does not exist, and a *symmetric maximum* such that, for given set of species, diversity is maximized when all species are equally frequent (Patil and Taillie, 1982). Each of the measures of diversity discussed in this chapter has these properties. A measure of species diversity also ideally should be nonparametric, applicable to any community independent of species abundance distribution, and statistically accurate, with small bias and sampling variance in samples of moderate size. The most popular nonparametric measures of species diversity are species richness, S, and measures based on Shannon information, H, and Simpson concentration, Λ, which account for relative abundance of species.

7.2.1 Species richness

Species richness is simply the number of species in a community or sample based on presence or absence. It depends equally on rare and common species, regardless of their relative abundance. Species richness is the most popular measure of species diversity because of its simplicity in data acquisition and analysis. However, it has the well known statistical weakness of a potentially large sampling bias, because rare species often will be absent even in large samples or intensive surveys.

Let S be the actual number of species in a community composed of a very large (effectively infinite) number of individuals. Although S is independent of the relative abundance of species in the community, the statistical properties of estimates of S

depend on species frequencies. Denoting the frequency of species i as p_i and using a carat to indicate a sample value, $\hat{S}$ is the number of species present in a random sample of n individuals from a community. Species i will be absent from the sample if none of the sampled individuals are from this species, which occurs with probability $(1 - p_i)^n$. The expected number of species in the sample is therefore

$$E[\hat{S}] = S - \sum_{i=1}^{S}(1 - p_i)^n \qquad 7.3a$$

(Grassle and Smith, 1976). Species i is likely to be present in the sample only if $np_i > 1$. Hence in a highly diverse community the observed number of species may have a large bias, because rare species often will be missing even from large samples, as shown in Figure 7.2.

Classical species–area curves describe the increase in observed numbers of species as a function of increasing area (Huston, 1994; Rosenzweig, 1995). This results from two factors. First, sampling bias decreases the number of species observed in small areas with limited sample size (eq. 7.3a). Second, larger areas tend to encompass greater geographic heterogeneity in community composition because of increased habitat heterogeneity and because local populations at more distant localities experience more independent environmental stochasticity.

The sampling variance of observed species richness can be derived from the variance of a sum of random variables representing either presence (1) or absence (0) of each species in a random sample of n individuals from the community,

$$\text{Var}[\hat{S}] = \sum_{i=1}^{S}(1 - p_i)^n[1 - (1 - p_i)^n]$$

$$+ 2\sum_{i=1}^{S}\sum_{j>i}^{S}[(1 - p_i - p_j)^n - (1 - p_i)^n(1 - p_j)^n]. \qquad 7.3b$$

(Strömgren *et al.*, 1973; corrected in Lande, 1996). The first (single) summation contains the variances of presence versus absence of each species. The second (double) summation contains the covariances of presence versus absence for pairs of species, which are always negative.

7.2.2 Shannon information

The Shannon information measure of diversity, although popular, has a rather tenuous foundation in ecological theory, as noted by Pielou (1966, 1969), Hurlbert (1971), and May (1975). Based on the frequency p_i of the ith species in the community the average information per individual in the community is

$$H = -\sum_{i=1}^{S} p_i \ln p_i \qquad 7.4a$$

(Shannon and Weaver, 1962). Shannon information depends most on species of intermediate abundance. This measure achieves its maximum value ln S when all species in a community are equally abundant. The exponential of Shannon information is often used, which has a maximum of S.

In a community with many species, Shannon information may have a substantial bias of unknown magnitude as we now show. The estimated information $\hat{H}$ is calculated from formula 7.4a using the observed species frequencies in the sample, $\hat{p}_i$. In a large random sample of n individuals the expected value and sampling variance of $\hat{H}$ are

$$E[\hat{H}] \cong H - \frac{S-1}{2n}$$ 7.4b

$$Var[\hat{H}] \cong \frac{1}{n} \left[\sum_{i=1}^{S} p_i (\ln p_i)^2 - H^2 \right]$$ 7.4c

(Pielou, 1966; Hutcheson, 1970; Bowman *et al.*, 1971). The bias in $\hat{H}$ depends on the actual number of species S which is often unknown. Thus an unbiased estimator of Shannon information generally does not exist. Accurate estimation of H requires large samples, much greater than the unknown actual number of species in the community, $n \gg S/2$, as indicated in Figure 7.3.

7.2.3 Simpson diversity

The probability that two randomly chosen individuals from a community are the same species is

$$\Lambda = \sum_{i=1}^{S} p_i^2$$ 7.5a

which is termed the 'concentration' (Simpson, 1949). The complement of Simpson concentration is the probability that two random individuals are different species, $1 - \Lambda$. This a useful measure of diversity termed the Simpson diversity (Pielou, 1969; Lande, 1996) or Gini index, which has a long history (Good, 1982). Simpson diversity depends most on the common species and achieves its maximum value $1 - 1/S$ when all species in a community are equally abundant. The inverse of Simpson concentration, $1/\Lambda$, has a maximum of S, and is often employed to measure diversity (Peet, 1974; May, 1975).

An analogy can be drawn between diversity measures, which quantify the variety of categorical entities (species) among individuals, and variance, which quantifies the dispersion of a metrical property among individuals (Pielou, 1975; Patil and Taillie, 1982). Simpson diversity can be expressed precisely as a variance by representing individuals in a community as points on an S-dimensional graph with individuals of species i located at one unit along the ith axis. The total variance per individual in the community is then $\sum_{i=1}^{S} p_i(1 - p_i) = 1 - \Lambda$ (Lande, 1996). Because this is a variance it follows that an estimate of Simpson diversity, $1 - \hat{\Lambda}$, calculated using the

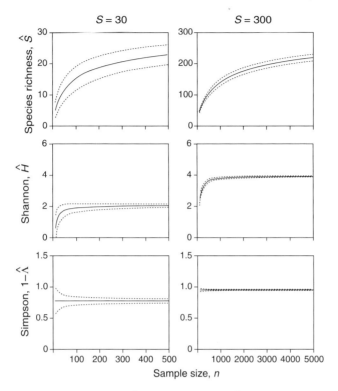

Fig. 7.3 Estimated species richness, $\hat{S}$, Shannon diversity, $\hat{H}$, and Simpson diversity, $1 - \hat{\Lambda}$, as a function of sample size, n, for hypothetical communities in which the species abundance distribution is lognormal and the standard deviation of the natural logarithm of species frequencies takes a typical value of $\sigma = 2$. Solid lines give the mean diversity, and dashed lines show approximate 95% confidence intervals ($\pm$ two standard deviations). In communities with a total of $S = 30$ or 300 species the frequency of the most abundant species is 42.6% or 17.4%, respectively. (From Lande, 1996.)

observed species frequencies, $\hat{p}_i$, in a random sample of n individuals has expected value $E[1 - \hat{\Lambda}] = \left(1 - 1/n\right)(1 - \Lambda)$ (Stuart and Ord, 1994). Thus an unbiased estimator of Simpson diversity is

$$1 - \tilde{\Lambda} = \left(\frac{n}{n-1}\right)(1 - \hat{\Lambda}) \qquad 7.5b$$

which has expected value equal to the actual Simpson diversity in the community, $E[1 - \tilde{\Lambda}] = 1 - \Lambda$. The approximate variance of the unbiased estimator in large samples is (Simpson, 1949; Lande, 1996)

$$\mathrm{Var}[1 - \tilde{\Lambda}] \cong \frac{4}{n}\left[\sum_{i=1}^{S} p_i^3 - \Lambda^2\right]. \qquad 7.5c$$

Sampling properties of the three commonly used nonparametric measures of species diversity are compared in Figure 7.3 for communities with low and high species richness and lognormal species abundance distributions. To summarize the main results, observed species richness can greatly underestimate the actual number of species even in large samples from a highly diverse community. Confidence intervals for observed species richness also may be rather wide. Shannon information has a substantial bias unless the sample size is much greater than the actual number of species, which, however, is generally unknown in highly diverse communities. Confidence intervals for Shannon information are much narrower than for species richness. The unbiased estimator of Simpson diversity, $1 - \tilde{\Lambda}$, is not only accurate, but also has narrow confidence intervals in samples of moderate size.

7.3 Species accumulation, rarefaction, and extrapolation

7.3.1 Species accumulation

The increase in observed species richness with increasing cumulative sample size, in the actual order that individuals are sampled in time, is called a species accumulation curve. This indicates roughly whether cumulative sampling is close to finding all species in the community, depending on whether the curve has leveled off at the largest sample sizes. Species accumulation curves also are used to control for differences in sample size in comparisons of species richness between areas. For example, DeVries *et al.* (1997) plotted species accumulation curves for different spatial segments of the fruit-feeding butterfly community in Amazonian Ecuador, as shown in Figure 7.4. This suggests that at any given sample size the canopy has higher species richness than the understory, and that forest edge has substantially lower species richness than other forest habitats. However, a species accumulation curve inevitably confounds increasing sample size with temporal and/or spatial components of species richness depending on how sample size is increased. Because of this and the unique order in which individuals are sampled, species accumulation curves are not amenable to statistical analysis unless the temporal and spatial components of diversity are ignored.

7.3.2 Rarefaction

The large bias in observed species richness, depending on sample size (eq. 7.3a), implies that it is statistically invalid to compare directly species richness in samples of different size. A method designed to equalize sample sizes from two communities is rarefaction (Sanders, 1968). This involves random sampling without replacement from the larger sample to create a random subset of the same size as the smaller sample. Consider a collection totalling n_T individuals of S species, with n_i individuals of the ith species. Let S_n denote the number of species in a random subset of

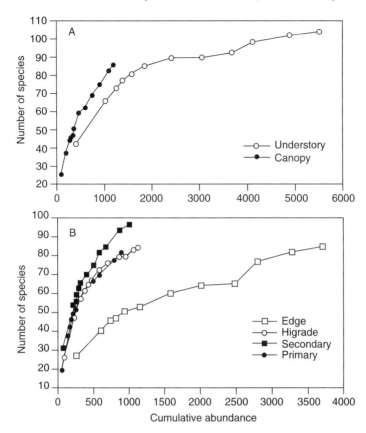

Fig. 7.4 Species accumulation curves showing observed species richness versus cumulative individual abundance through time in the order of sampling. (A) Canopy and understory. (B) Four forest habitats: primary (undisturbed), secondary (logged and regenerated), higrade (partially logged), and edge (adjacent to land cleared for pasture). (From DeVries *et al.*, 1997.)

size n. The probability that species i is absent from this subset, using combinatorial notation, is

$$\chi_n(n_i) = \binom{n_T - n_i}{n} \Big/ \binom{n_T}{n} \quad \text{where} \quad \binom{n_T}{n} = \frac{n_T!}{(n_T - n)!n!}.$$

The mean species richness in random subsets of size n is

$$E[S_n] = S - \sum_{i=1}^{S} \chi_n(n_i) \qquad \text{7.6a}$$

(Hurlbert, 1971). A graph of $E[S_n]$ plotted against sample size n is called a rarefaction curve. The variance of species richness in random subsets of size n is

$$\text{Var}[S_n] = \sum_{i=1}^{S} \chi_n(n_i)[1 - \chi_n(n_i)] + 2\sum_{i=1}^{S}\sum_{j>i}^{S}[(\chi_n(n_i + n_j) - \chi_n(n_i)\chi_n(n_j)].$$

7.6b

where $\chi_n(n_i + n_j)$ is the probability that both species i and j are absent from a random subset of n individuals (Heck *et al.*, 1975). If n is reasonably large, species richness usually will be approximated normally distributed so that the 95% confidence interval can be approximated by $E[S_n] \pm 2\sqrt{\text{Var}[S_n]}$. As n_T approaches infinity, equations 7.6a,b become identical to 7.3a,b.

DeVries *et al.* (1997) developed a simple method of testing for differences in species richness among subdivisions of a community of taxonomically related species in the same ecological guild. They applied rarefaction to the total sample from the fruit-feeding nymphalid butterfly community in Figure 7.1 to account for differences in sample size in assessing whether subdivisions of the community had different species richness than the whole community. Figure 7.5 shows the rarefaction curve for the total sample with confidence intervals. Parts of the community are plotted on this graph as symbols located at their observed species richness and sample size. This reveals that the numbers of species sampled in understory and forest edge habitats are substantially and significantly less than expected in a random sample of the same size

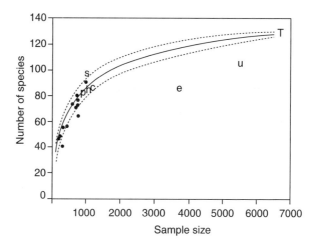

Fig. 7.5 Rarefaction curve (solid curve) and approximate 95% confidence intervals (dashed curves) for the total sample of fruit-feeding nymphalid butterflies (Fig. 7.1) compared with observed species richness in subdivisions of the community along dimensions of height, habitat and time. T = total community. Vertical stratification: c = canopy, u = understory. Horizontal forest types: p = primary, s = secondary, h = higrade, e = edge. Temporal: • = months. (From DeVries *et al.*, 1997.)

from the community as a whole, while the number of species sampled from other spatial and temporal subdivisions did not differ significantly from that a random sample of equal size from the total community. Samples from parts of the community that sum to the total community must on average lie below the rarefaction curve for the total sample if there is significant heterogeneity in species composition among the parts. Chi-squared tests for homogeneity of species abundance distributions revealed significant heterogeneity between the canopy and understory, as well as among habitat types and monthly samples (DeVries *et al.*, 1997). More species were observed in the understory than in the canopy, but many more individuals were sampled in the understory despite the same sampling effort with standardized traps in both habitats; thus their species richness cannot be directly compared. The canopy sample lies well within the confidence intervals for the rarefaction curve for the total community, whereas the understory sample is very significantly below the rarefaction curve. This suggests that, although more species were observed in the larger sample collected in the understory, after correcting for sample size the canopy has substantially higher species richness than the understory. However, sampling the same butterfly community at another site 160 km away, indicated that species richness in the canopy did not differ significantly from that in the understory (DeVries *et al.*, 1999).

7.3.3 Extrapolation

Various methods have been devised to extrapolate actual species richness from samples (Palmer, 1990, 1991; Colwell and Coddington, 1994). A random sample of individuals from a community can be used to fit an assumed form of species abundance distribution to estimate the number of species missing from a sample. For highly diverse communities a rather large sample is required to estimate accurately the parameters of the species abundance distribution (Bulmer, 1974; Pielou, 1974; Kempton, 1979). For example, if the true species abundance distribution is lognormal but the sample is not sufficiently large to reveal the mode of the distribution to the right of the truncation point, it usually is not possible to distinguish statistically the lognormal distribution from Fisher's logseries distribution, which assumes an infinite number of species in the community (see Fig. 7.2 and Kempton, 1979). For this reason, Fisher's logseries distribution (after eq. 7.2) frequently is employed to describe insect communities containing hundreds of species. Parametric extrapolation of actual species richness for highly diverse communities appears to be of limited utility because it generally is impractical to test empirically, which would require astronomically large sample sizes.

Even 'nonparametric' estimators of actual species richness (Chao and Lee, 1992; Bunge and Fitzpatrick, 1993) make implicit assumptions about species abundance distributions, as in principle there could be an enormous number of extremely rare species not present in the sample. No unbiased estimator of species richness exists except in the unusual case when the sample size is at least as large as the population size of the most abundant species (Goodman, 1949; Engen, 1978). Without regard to the species abundance distribution the only aspect of unobserved species that can be accurately estimated is their total frequency in a community, from the number of 'singletons' (species represented by one individual in the sample) divided by the sample size (Good, 1953; Engen, 1978). Actual species richness cannot be accurately extrapolated by any method using small samples containing a minority of species in a community (Engen, 1978; Chao and Lee, 1992; Bunge and Fitzpatrick, 1993). In a speciose community the actual number of species can be accurately estimated only by exhaustive sampling or by extrapolation using an assumption about the form of the species abundance distribution.

7.4 Rapid assessment of diversity

Widespread degradation and destruction of natural habitats has motivated the use of rapid surveys of species diversity in ecological monitoring and conservation (Roberts, 1991; Beattie, 1993; Conservation International, 2002*b*). When time is short and resources are limited this necessarily involves using small samples to rank species diversity in different areas (Beattie, 1993; Conservation International 2002*b*). The measure of species diversity typically employed in rapid assessments is observed species richness, which in small samples from highly diversity communities has a large bias that depends strongly on sample size (eq. 7.3a, Fig. 7.3). Observed species richness is not a reliable measure to rank accurately actual species richness in different areas, unless the areas have been exhaustively surveyed to compile complete species lists. This applies even with standardized sampling methods advocated by many workers (Spellerberg, 1991; Stork and Samways, 1995) and especially when data on species abundances are lacking.

In this context it is of considerable interest that observed species richness in small samples more accurately reflects the Simpson diversity rather than the actual species richness of a community. This can be seen by noting that the expected initial slope of a species accumulation curve is the expected number of species observed in a sample of two individuals minus that in a sample of one individual. A sample of one individual always contains one species. The expected number of species in a random sample of two individuals is $2 - \Lambda$, the sum of 1 for the first individual plus the probability that the second individual is a different species from the first, the Simpson diversity $1 - \Lambda$ (eqs 7.5a,b). Hence the expected initial slope of the species abundance distribution equals the Simpson diversity $1 - \Lambda$. The initial slope of a species accumulation curve has little or no relation to its asymptote at very large sample sizes, the actual species richness in the community.

Observed species richness in small samples will tend to rank incorrectly the actual species richness of two areas when their species accumulation curves intersect, which is expected to occur when the area with the higher Simpson diversity has a lower actual species richness. This may often happen when comparing areas that differ in degree of natural or artificial disturbance. An intermediate magnitude of disturbance tends to increase species richness, while more severe disturbance releases some species

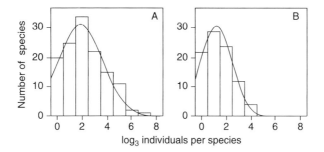

Fig. 7.6 Species abundance distributions of fruit-feeding nymphalid butterflies in two low-land Amazonian forest sites covering 400 ha sampled monthly in canopy and understory over 1 year. (A) Disturbed site at Jatun Sacha, Ecuador with sample size 6,690 individuals, as in Figure 7.1. (B) Intact site at La Selva Lodge, Ecuador with sample size 883 individuals. Lognormal distributions fitted by the method of Pielou (1975). Rarefaction of the larger sample to the size of the smaller sample showed that the twofold difference in variances of the fitted lognormal distributions is statistically significant ($P < 0.01$ in 10,000 rarefied samples). (From DeVries *et al.*, 1999; Lande *et al.*, 2000.)

to high abundance and causes other species to become rare, thereby increasing the variance of the species abundance distribution (Patrick, 1963; Kempton and Taylor, 1974; Connell, 1978; Preston, 1980) and decreasing Simpson diversity. Habitat heterogeneity within the mean individual dispersal distance typical of species in the community also increases species richness and the variance of species abundances (Connell, 1978; Preston, 1980). Thus intersection of species accumulation curves is especially likely to occur when one sampling area consists of undisturbed habitat and the other area encompasses a broad range of disturbance regimes.

Species abundance distributions at disturbed and intact sites are compared for fruit-feeding butterfly communities in Amazonian Ecuador in Figure 7.6, showing that the former has a significantly larger variance. The fitted lognormal distributions enable extrapolation of the number and frequencies of unsampled species to estimate the actual species richness for each community and to calculate their expected species accumulation curves (eq. 7.3a) neglecting the small seasonal component of species richness (see Fig. 7.5). The disturbed site has a higher estimated actual species richness and a lower Simpson diversity than the intact site (see Table 7.1), so that their expected species accumulation curves intersect as in Figure 7.7. One species, *Cissia penelope*, that was quite rare in the intact community, at a frequency of 0.23%, was released by the presence of artificial forest edge adjacent to a pasture to become the most abundant species at the disturbed site, comprising 24% of the community. In the intact site, which included only natural lakeside edge, the most abundant species, *Nessaea hewitsoni*, composed 12% of the community (DeVries *et al.*, 1997, 1999). This difference in frequencies of the most abundant species largely accounts for the difference in Simpson diversity between the communities. A large sample from each community is required for the observed species richness

Table 7.1 Sample sizes for confidently ranking diversity measures in the two butterfly communities in Figure 7.6. (From Lande *et al.*, 2000)

Diversity measure	Estimates		Minimum sample[1]
	Disturbed	Intact	
Species richness[2]	142	101	1,801
Simpson diversity[3]	0.917	0.954	81

[1] Minimum size of equal samples from the two communities required for the ranking of observed diversity to correctly represent actual diversity in the communities with 95% confidence.
[2] Actual species richness estimated assuming lognormal species abundance distributions.
[3] Simpson diversity estimated directly from total samples in Figure 7.6, using the unbiased estimator (eq. 7.5b).

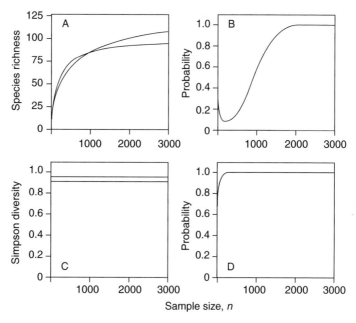

Fig. 7.7 Ranking diversity measures in the two butterfly communities in Figure 7.6, as a function of the size of random samples from each community, ignoring the seasonal component of diversity at both sites (DeVries *et al.*, 1997, 1999). (A) Expected species accumulation curves (expected species richness from eq. 7.3b). (B) Probability that the ranking of observed species richness correctly represents actual species richness in the two communities. (C) Expected Simpson diversity; upper line for intact community, lower line for disturbed community. (D) Probability that the ranking of estimated Simpson diversity correctly represents actual Simpson diversity in the two communities. (From Lande *et al.*, 2000.)

to have a high probability of correctly ranking the actual species richness in the disturbed and intact areas, whereas modest samples are sufficient for the estimated Simpson diversity to correctly rank the actual Simpson diversity with high probability (Table 7.1). Rapid assessments should therefore make use of Simpson diversity as well as observed species richness when comparing species diversity in different areas.

7.5 Partitioning diversity into additive components

The general concept of biodiversity is rather vague, as should be evident from our discussion of different measures of species diversity. Although it is intuitively obvious that human activities are greatly reducing biodiversity, this can vary from depleting the number of species at a particular place and time, to homogenizing species composition among localities. For example, Harrison (1993) showed that the lists of bird species on Pacific islands tend to be homogenized by anthropogenic extinctions of native endemic species and widespread introductions of exotic species, with a net reduction of total species richness. Conservation efforts may be misguided unless we pay careful attention to precisely how human activities alter species diversity in space and time. To generate these insights we need to partition species diversity into components, much like an experimentalist uses Analysis of Variance to partition a variance into components associated with different factors or causes. Here we develop a method for partitioning any measure of species diversity into additive components. This approach has major advantages over the traditional multiplicative definition of components of diversity.

Whittaker (1960, 1972) defined for species richness the important concepts of α-diversity within localities, γ-diversity in the total community, and β-diversity or proportional species turnover between locations or along an environmental gradient. Whittaker's definition of β-diversity entails a multiplicative partition of diversity, $\gamma = \alpha\beta$ so that $\beta = \gamma/\alpha$ is dimensionless and cannot be directly compared with α and γ components. This also makes it difficult to extend the analysis to a hierarchically structured community with multiple levels of subdivision and is not conducive to statistical analysis. We demonstrate below that, for any measure of diversity satisfying certain requirements, the total community diversity can be partitioned into additive components within and among subdivisions of the community, such that total diversity, $\gamma = \alpha + \beta$, and its components α and β have the same units and can be compared directly. An additive partition of diversity therefore appears more natural than the multiplicative partition used by Whittaker and many subsequent ecologists (e.g. MacArthur, 1965, 1972; Pielou, 1975; Routledge, 1977; Magurran, 1988), and is more amenable to hierarchical sampling and statistical analysis.

Levins (1968; pp. 49–50) first applied an additive partition of diversity using Shannon information to compare temporal and spatial components of species diversity in *Drosophila* communities at different latitudes. Lewontin (1972) developed and applied this approach to partition the Shannon information measure of genetic diversity within and among human races and populations. This prompted Nei (1973) to develop methods for partitioning genetic heterozygosity, which is equivalent to Simpson diversity, to measure genetic differentiation among populations, a method that is now commonly used by population geneticists to analyze the geographic structure of populations. Allan (1975a,b) explained and applied the method to partition additively the Shannon information measure of species diversity. Applications of additive partition of species diversity then languished until reviewed by Lande (1996) and applied by DeVries *et al.* (1997) and other authors (DeVries *et al.*, 1999; Wagner *et al.*, 2000; DeVries and Walla, 2001; Fournier and Loreau, 2001).

Lewontin (1972) proposed that a desirable property for any diversity measure is that the total diversity in a community should be greater than or equal to the (weighted) average diversity within subdivisions of the community. This property permits a measure of species diversity in a community to be partitioned into additive non-negative components within and among subdivisions, as we now show. Let D_T be a measure of the total species diversity in a community, computed using the weighted average frequency of each species among parts of the community. The weights may be arbitrarily chosen. For example, equivalent weights would reflect equal importance placed on different parts of the community. Weighting in proportion to actual abundance or sample size for the different parts can be done simply by pooling raw data for the total community or sample. Denote the diversity within the jth part of the community as D_j and let the proportional weight of this part be q_j such that $\sum_j q_j = 1$. A diversity measure is *strictly concave* (upwards) when

$$D_T \geq \sum_j q_j D_j \qquad \text{7.7a}$$

with equality only when all parts of the community have identical species composition.

Species richness, S, is strictly concave. A continuous measure of diversity, D, based on species frequencies, p_i, is strictly concave if and only if its matrix of second derivatives (with matrix elements $\partial^2 D/\partial p_i \partial p_j$) *is negative definite* (Marcus and Minc, 1964). Shannon information, H, is strictly concave (Aczél and Daróczy, 1975, p. 35). Simpson concentration, Λ, is strictly convex, hence Simpson diversity, $1 - \Lambda$, is strictly concave. However, the commonly used inverse Simpson concentration measure of diversity, $1/\Lambda$, is not concave in general (Patil and Taillie, 1982; Lande, 1996). Thus in some cases using the inverse Simpson concentration, the total diversity in a community may be less than the average diversity within its parts, implying the possibility of a negative β-diversity among the parts. This always occurs when a community of two species is subdivided into two parts each of which has the most common species in frequency exceeding $(1 + \sqrt{3})/2 \cong 0.789$, as in the numerical example in Table 7.2. (For a multispecies example see Lande, 1996.)

For any diversity measure that is always non-negative and strictly concave the total diversity in a community can be additively partitioned into non-negative components within and among subdivisions of the community,

$$D_T = \bar{D}_{within} + D_{among} \quad \text{where } \bar{D}_{within} = \sum_j q_j D_j. \qquad \text{7.7b}$$

Table 7.2 An example in which the inverse Simpson concentration measure of diversity, $1/\Lambda$, violates concavity, so that diversity of the total community is less than the average diversity within its parts. Pooled parts have equal weights

	Species frequencies		
	Part 1	Part 2	Pooled
p_1	0.8	1.0	0.9
p_2	0.2	0.0	0.1
$1/\Lambda$	1.4706	1.0000	1.2195
$\frac{1}{2}(1/\Lambda_1 + 1/\Lambda_2)$			1.2353

In practice the diversity among subdivisions of a community can be calculated by subtracting the weighted average diversity within the parts from the total community diversity, $D_{among} = D_T - \bar{D}_{within}$ where the total diversity is calculated from the weighted average species frequencies among the parts.

A natural measure of similarity among subdivisions of a community is $\psi_D = \bar{D}_{within}/D_T$, which ranges between 0 and 1. This reveals for species richness that Whittaker's measure of β-diversity is not actually a diversity, but rather the inverse of community similarity in species composition.

Additive partition of diversity can be extended to multiple levels of subdivision of a community, provided that the subdivisions are hierarchically structured. Consider two hierarchical levels, with level 1 being the lower (smaller) subdivisions and level 2 the higher (larger) subdivisions. Moving down the hierarchy, first notice (from eq. 7.7b) that the total community diversity can be partitioned as $\gamma = \alpha_2 + \beta_2$. Diversity also can be partitioned within and among level 1 units within the kth unit of level 2, so that $\alpha_{k2} = \alpha_{k1} + \beta_{k1}$, and taking weighted averages of these values across the larger units yields $\alpha_2 = \alpha_1 + \beta_1$. Combining this with the first partition produces $\gamma = \alpha_1 + \beta_1 + \beta_2$ (Patil and Taillie, 1982; Wagner *et al.*, 2000; Fournier and Loreau, 2001).

7.5.1 Randomization tests of significance

Solow (1993) suggested a randomization test (Manly, 1991) to assess the statistical significance of an observed difference in samples from different communities or parts of a single community. First, a test statistic measuring the difference between the samples is computed. In the present context this measure is β-diversity. Then all individuals in the samples to be compared are pooled and randomly reassigned to the samples, preserving the original sample sizes to create a new data set of the same form as the original, and recomputing the test statistic. This randomization procedure is performed a large number of times to create a distribution of the test statistic under the null hypothesis that the samples were randomly drawn from a homogeneous community. The probability that the distribution of the test statistic under the null hypothesis exceeds the test statistic for the original data gives the significance level (*P* value) of the test statistic for the original data. The validity of this test requires that individuals are randomly sampled within each part of the community. It is not necessary that all species are included in the sample.

The randomization test for significance of β-diversity can be generalized to hierarchical sampling schemes by using a procedure involving restricted randomization at a given level of the hierarchy. The test statistic measuring differences among sampling units of a community at level k in the hierarchy is the β_k component of diversity. The null hypothesis for this test is that the sampling units at the next lowest $(k - 1)$ level of subdivision of the community are randomly distributed among the sampling units at level k. The distribution of the test statistic under the null hypothesis is constructed by randomizing only the level $k - 1$ units among level k units within each $k + 1$ level unit, preserving the original sample sizes within each of the k level units and maintaining the original data composition within each of the $k - 1$ level units. This is analogous to a nested Analysis of Variance in which the significance of variation at a given level is tested against variation at the next lower level in the hierarchy (Scheffé, 1959). Consider for example the hierarchical sampling design in Table 7.4. To test the significance of β_2-diversity among stands within sites would involve randomizing trees among stands within sites, maintaining the original composition of individuals sampled within trees and preserving the number of trees within stands.

Complete randomization of individuals across all of the lowest level sampling units in the entire community produces a partitioning of diversity corresponding to the null hypothesis of a completely homogeneous community. This will be referred to as the null model of community structure (Gotelli and Graves, 1996).

7.5.2 Spatial components of diversity

De Vries *et al.* (1997) partitioned measures of tropical butterfly species diversity within and among four forest habitats by pooling monthly samples obtained over 1 year from a site in lowland Amazonian Ecuador. Significant β-diversity was indicated by Chi-squared tests for homogeneity of species abundance distributions among habitats. The additive partition of diversity measures in Table 7.3 reveals that among forest habitat types there occurs about 33% of species richness, 14% of Shannon information,

Table 7.3 Additive partition of butterfly species diversity among four forest habitats (primary, secondary, higrade, and edge) in 6,690 individuals sampled over 1 year in Amazonian Ecuador, depicted in Figure 7.1 (DeVries *et al.*, 1997)

Diversity components	Species richness	Shannon information	Simpson diversity
Among habitats, β	43.03	0.466	0.060
Within habitats, α	86.97	2.910	0.857
Total, γ	130	3.376	0.917

Table 7.4 Additive partition of beetle species diversity among 8,662 individuals from 467 species in a hierarchical sample of deciduous forest in eastern U.S.A. Significance of diversity components was assessed by restricted randomization at a given level of the hierarchy. Also shown are the expected diversity components on the alternate null hypothesis of no community structure, H_1, obtained by complete randomization of individuals across all levels of the hierarchy. (From Crist *et al.*, 2002)

Diversity component	Species richness		Shannon info.		Simpson diversity	
	Observed	H_1	Observed	H_1	Observed	H_1
Among regions, β_4	137.7	90.2	0.26	0.03	0.012	0.0001
Among sites within regions, β_3	136.6**	124.8	0.38*	0.10	0.027	0.0004
Among stands within sites, β_2	101.5**	115.8	0.49	0.26	0.034	0.002
Among trees within stands, β_1	52.4	72.5	0.59	0.50	0.030***	0.007
Individuals within trees, α_1	38.7	63.7	2.86	3.69	0.868	0.962
Total, γ	467	467	4.58	4.58	0.971	0.971

Observed > Expected: * $P < 0.05$,** $P < 0.001$; Observed < Expected: *** $P < 0.001$.

and 6.5% of Simpson diversity. These results reflect that the rarest species represented by only a few individuals in the total sample cannot appear in all four habitats, and that common species occurred in similar frequencies in each habitat.

Crist *et al.* (2002) partitioned species diversity for beetles collected by insecticidal fogging of canopy tree crowns at four levels of hierarchical sampling of deciduous forest throughout the eastern U.S.A. There is a striking difference in the way that the three measures of diversity are partitioned in Table 7.4. Simpson diversity occurs mostly within trees, again reflecting that common species are consistently abundant across the range of the study. Restricted randomization tests show that the β_1 component of Simpson diversity among trees within stands is very highly significantly less than expected by chance, indicating that some ecological mechanism maintains the relative abundance of common species within stands. Species richness occurs mostly among higher levels of the hierarchy, because rare species represented by a few individuals tend to occur in different locations across the range. Restricted randomization tests show that the β_2 and β_3 components of species richness among stands within sites and among sites within regions are very highly significantly greater than expected by chance, indicating that the tendency for different rare species to occur in different localities is not just an artifact of their small sample sizes. Shannon information occurs mostly within trees but is distributed substantially across all higher levels in the hierarchy, reflecting variation in geographic distribution among species with intermediate frequencies. For each of the three measures, the observed partition of diversity differs greatly (and very highly significantly) from that under the null model of community structure obtained by complete randomization of individuals across all trees in the study (H_1).

Methods have been suggested for additive partitioning of diversity measures based on crossed factorial designs (Goodman, 1970; Alatalo and Alatalo, 1977; Patil

Table 7.5 Additive partition of butterfly species diversity among 5 years of monthly sampling at La Selva Lodge, Ecuador totalling 11,861 individuals (DeVries and Walla, 2001). Monthly samples were pooled within years and significance of the among year similarity was assessed by 10,000 randomizations of individuals among years (J. Veech, pers. com.)

Community similarity, $1 - \beta/\gamma$ or total diversity, γ	Species richness	Shannon information	Simpson diversity
Observed similarity	0.7359*	0.9687*	0.9849*
Expected similarity	0.7594	0.9931	0.9997
Total diversity	128	3.355	0.9337

Observed < Expected: * $P < 0.001$.

and Taillie, 1982; Rao, 1984), as often employed in factorial Analysis of Variance (Scheffé, 1959). For example, the interaction of temporal and spatial factors affecting diversity could be investigated if species abundances are sampled in the same set of areas through time. To the best of our knowledge this approach has not yet been applied to ecological data to analyze interactions of factors affecting species diversity.

7.5.3 Temporal components of diversity

Rosenzweig (1995) emphasized that relatively little is known about the temporal component of species diversity. The dearth of knowledge on this subject is due not only to previous methodological inadequacy, but also to the difficulty of conducting standardized sampling over biologically meaningful time scales much longer than the generation times of species in the community. This is most easily accomplished using species with short generations. For instance, Levins (1968) sampled *Drosophila* communities to show that the seasonal component of Shannon information is larger in the temperate zone than in the equatorial tropics.

DeVries and Walla (2001) sampled fruit-feeding nymphalid butterflies at a site in lowland Amazonian Ecuador (160 km from that of DeVries *et al.*, 1997) every month for 5 years. Pooling the monthly samples within years facilitates analysis of the temporal component of species diversity among years, as in Table 7.5, with community similarity among yearly samples defined as $1 - \beta/\gamma$, where β and γ represent the among year temporal component and the total diversity respectively (Lande, 1996). Significance of the among year component was assessed by randomizing individuals among years, from the proportion of randomizations in which the similarity exceeded that observed in the original data. For each of the diversity measures the observed similarity among years was very highly significantly less than expected under randomization, indicating a substantial temporal component of β diversity. DeVries and Walla (2001) reached a similar conclusion based on a Chi-squared test for heterogeneity of species abundance distributions among years.

7.6 Summary

Species abundance distributions in natural communities often can be described by Preston's lognormal model or by a Gamma distribution. With random sampling of individuals from a community, the corresponding sampling distributions of observed species abundance are the Poisson-lognormal and the negative binomial. Several other models of species abundance distributions are special cases, or can be closely approximated, by these distributions. Species diversity in a community is commonly described using a nonparametric index such as species richness, Shannon information, or Simpson diversity. These measures differ greatly in the accuracy with which they can be estimated from random samples of a community. Estimates of species richness may have a large downward bias because rare species are likely to be absent even from large samples. This bias depends strongly on sample size, so it is not valid to compare observed species richness directly using samples of different size. Rarefaction of a larger sample to the size of a smaller one can be used to compare observed species richness in samples of different size. For a speciose community the actual number of species can be accurately estimated only by exhaustive sampling or by extrapolation using an assumption about the form of the species abundance distribution. In rapid assessments of species diversity, different species abundance distributions in intact versus disturbed communities are likely to cause their species richness to be misranked unless the sample sizes are large, whereas even modest samples will correctly rank their Simpson diversity with high probability.

Any measure of species diversity should ideally be non-negative, with a symmetric maximum, and strictly concave. Strict concavity entails that the diversity in the total community is always at least as large as the (weighted) average diversity within subdivisions of the community. A non-negative strictly concave measure of species diversity can be partitioned into non-negative additive components, such that the total or γ-diversity equals the (weighted) average or α-diversity within subdivisions plus the β-diversity among subdivisions. Most of the commonly used nonparametric measures of species diversity are strictly concave, but not the inverse Simpson measure, which allows negative β-diversity. In comparison with Whittaker's multiplicative partitioning, additive partitioning of diversity has the advantage that all components of diversity have the same units and can be directly compared. Additive partitioning also facilitates analysis of hierarchical or factorial sampling of a community to investigate spatial, temporal, and ecological factors affecting diversity, similar to an Analysis of Variance. The statistical significance of components of species diversity can be tested by randomization and restricted randomization.

8

Community dynamics

Most of conservation biology deals with one species at a time. However, increasing attention is being devoted to the conservation and restoration of landscapes, communities and ecosystems and how their dynamics are altered by human activities (Edwards *et al.*, 1994; Scott *et al.*, 1996; Soulé and Terborg, 1999). One reason we see relatively little discussion of community dynamics in conservation is the absence of any guiding theory. A leading theory of community dynamics consists of neutral models based on the assumptions that interacting species in a single trophic guild, such as tropical canopy trees, are ecologically equivalent and their relative abundances change only by demographic stochasticity. Despite these simplistic assumptions, neutral models appear to explain some classical ecological patterns, including species abundance distributions and species–area curves (Caswell, 1976; Bell, 2001; Hubbell, 2001). In this chapter we develop a more general model of community dynamics that allows species to differ in their basic ecology and responses to environmental fluctuations. We investigate the extent to which species in a community differ consistently in abundance because of differences in their ecology, and explore how environmental stochasticity influences community composition in space and time.

We apply this theory to analyze the dynamics of a tropical butterfly community sampled intensively through a limited space and time, emphasizing the advantages of studying taxa with short generations to facilitate sampling on temporal scales longer than the generation times of species in the community. Much of the recent application of neutral community models has concerned perennial plants, such as tropical trees, which have long generations (Bell, 2001; Hubbell, 2001; Kelly and Bowler, 2002). Butterfly generations are on the order of months, so that several years of sampling permit estimation of the strength of intraspecific density dependence and the temporal as well as spatial autocorrelation of the community composition. Our analysis shows that the majority of total variance in species abundance is attributable to ecological heterogeneity among species and to both general and species-specific responses to environmental variability acting over short temporal and spatial scales. Such heterogeneity among species may play a crucial part in enhancing the resilience of trophic structure and stability of ecosystem function during periods of major environmental change, such as global warming, increasing the chance that a complex community harbors some (perhaps currently rare) species that will thrive in a substantially altered environment (Patrick, 1963; Azuma *et al.*, 1997; DeVries *et al.*, 1997, 1999; Finlay *et al.*, 1997; Tilman *et al.*, 2001; Kinzig *et al.*, 2002).

8.1 Simple ecosystems

Realistic models of simple ecosystems containing a few dominant species with strong interactions can be formulated and fit to empirical observations of deterministic and stochastic factors affecting population dynamics. In classic laboratory experiments on competition between small populations of two species of flower beetle, *Tribolium castaneum* and *T. confusum*, Park (1957) demonstrated indeterminate outcomes, and Bartlett (1960) analyzed this as resulting from demographic stochasticity.

More recently, the impact of environmental stochasticity has been analyzed in simple natural ecosystems. For example, Caughley and Gunn (1993) modeled the dependence of the Red Kangaroo on their plant forage and the influence of rainfall on density-dependent plant population growth rate in an Australian desert ecosystem, using experimental data on food intake rate by kangaroos as a function of plant density, along with time series data on rainfall, plants, and kangaroos. The experimental data, combined with the high variability of this simple ecosystem, facilitated a description of kangaroo population growth rate as a nonlinear function of plant density, and they found no evidence of additional density dependence due to competition among kangaroos for space or territory. Although the environmental stochasticity in rainfall has no significant autocorrelation across years, the internal dynamics of this system produces long-term fluctuations and apparent trends lasting decades in the kangaroo population. Realistic models also have been constructed for the rainfall-dependent ecosystem of pasture, rabbits, and foxes in arid regions of Australia (Pech and Hood, 1998; Davis *et al.*, 2002). Stochastic simulations of these models display many of the observed features of these simple ecosystems, including sporadic outbreaks of herbivores. Similar models have been developed for predicting plagues of house mice in Australia (Pech *et al.*, 1999). Such detailed descriptions of ecosystem dynamics usually are precluded for complex ecosystems by the lack of sufficient information to estimate accurately the effects of environmental variables on each species as well as the ecological interactions among species.

Few analytical solutions exist for stochastic models of interacting species. A linearized model of an ecosystem around a stable equilibrium point can be derived by ignoring demographic stochasticity and assuming small environmental stochasticity, producing a diffusion model with constant infinitesimal variances and covariances. From this one can derive a stationary distribution of population sizes that is approximately multivariate normal, and make theoretical inferences about community stability and species coexistence in a fluctuating environment (May, 1974; and see Abrams, 1976; Turelli, 1978). More general nonlinear models have been proposed for numerical calculation of stationary distributions and rates of extinction for particular species in a community (Gilpin, 1974; Ludwig, 1975), still using special forms of stochasticity that are unrealistic for small populations. The most general diffusion model for extinction of species in a community requires extensive numerical analysis (Naeh *et al.*, 1990) and therefore has few advantages over more straightforward simulations, especially when more than a few species are involved.

8.2 Complex ecosystems and communities

Following the classical theoretical studies of Volterra (1928, 1931) and laboratory competition experiments of Gause (1934) on yeast and paramecia, it was long believed that in a constant environment two species cannot stably coexist in the same ecological niche, using exactly the same resources. This principle of competitive exclusion was extended to show that S species cannot coexist at a stable equilibrium on less than S different resources or limiting factors, based on general differential equation models for the population sizes of interacting species (Volterra, 1931; Levin, 1970). These results have strongly influenced ecological thinking (Hutchinson, 1957, 1959). Many detailed biogeographic and ecological studies have shown that closely related (congeneric) species either do not coexist in the same area (often having disjunct or adjacent geographic ranges), or are not ecologically equivalent when they do coexist (Huston, 1994; Brown and Lomolino, 1998).

More recent theoretical results have demonstrated that stable oscillations and chaos generated by nonlinear interspecific interactions can allow S species to coexist on fewer than S resources in constant environment (Armstrong and McGehee, 1980; Huisman and Weissing, 1999). However, little evidence exists to suggest that this is of much importance in complex natural communities (Schippers et al., 2001).

A persistant problem is to explain why complex ecosystems, such as tropical rainforests, contain communities with large numbers of taxonomically related and/or ecologically similar species. This is especially perplexing for communities that seem relatively homogeneous in space and in time, such as oceanic plankton (Hutchinson, 1961), and tropical rainforest trees and insects (Huston, 1994; Rosenzweig, 1995; Brown and Lomolino, 1998). One class of explanations consists of equilibrium theories. Many ecological niches may arise from heterogeneity in resources occurring in a geographic mosaic and/or along environmental gradients with species specialized to different combinations of resources (Tilman and Pacala, 1993). For tropical trees, minor topographic features can affect soil moisture and insolation, and soil composition and nutrients also may vary over short distances (Huston, 1994, ch. 14). The diversity of tropical herbivorous insects could then reflect specialization to a diversity of tropical plants and so on to higher trophic levels (Leigh, 1999, ch. 8). Another potentially important factor promoting equilibrium coexistence of species is intraspecific density dependence that limits common species and favors rare species. This may occur for a variety of reasons, including stronger competition between conspecifics than heterospecifics, e.g. by behavioral interactions such as territoriality in animal species. Species-specific pathogens and parasites, which occur throughout the animal and plant kingdoms, exert specific density limitation on their host populations through classical epidemiological mechanisms in which host crowding facilitates pathogen and parasite transmission (Poulin, 1998), thus favoring rare species. The Janzen–Connell hypothesis for the maintenance of species diversity in tropical trees and other communities concerns species-specific or locally adapted pathogen and herbivore limitation acting within progeny groups (from individual parents) at early life stages with limited dispersal, such as seedlings. This mechanism favors reproduction by individual parents somewhat isolated from conspecifics and favors offspring that disperse far from their parents (Janzen, 1970; Connell, 1971); it is increasingly supported by empirical evidence (Leigh, 1999, ch. 8; Webb and Peart, 1999; Harms *et al.*, 2000; Wright, 2002).

An alternate class of explanations consists of nonequilibrium theories that necessarily involve temporal dynamics of species abundances (Chesson and Case, 1986). As explained below, in a constant environment evolution may produce new ecologically similar or equivalent species faster than they become extinct through competition or chance events; and short-term environmental fluctuations also can maintain species diversity. Recent models demonstrate that in a fluctuating environment two or more species can stably coexist on a single resource (Levins, 1979; Chesson and Warner, 1981; Abrams, 1984; Chesson, 2000). Deterministically or stochastically fluctuating environments allow different species to coexist in the same ecological niche using exactly the same resources by specializing on different resource levels, so that each species has (on average) a competitive advantage when rare. This can also be explained by each species doing relatively well some of the time (in certain environments), or as a 'storage effect' for species with long-lived adults in which populations that usually tend to decline are maintained by occasional years of high recruitment occurring asynchronously for different species (Chesson and Warner, 1981; Chesson and Huntly, 1997; Kelly and Bowler, 2002).

Caswell (1976) introduced neutral models of community dynamics by drawing an analogy between species and selectively neutral alleles at a genetic locus, with speciation occurring like random mutation at a constant rate, and extinction occurring by demographic stochasticity. This allows numerous species to coexist transiently in a single ecological niche. Caswell (1976), Hubbell (1995, 1997, 2001), and Bell (2000, 2001) developed this approach in large part by borrowing results from the neutral theory of molecular evolution developed by Kimura (1983) and others, applying and extending it to classical ecological patterns such as species abundance distributions and species–area curves. Neutral community models incorporate the key assumptions that all species in a community are ecologically identical, with their relative abundance changing only by demographic stochasticity, analogous to random genetic drift of allele frequencies. Hubbell allows general environmental variation to produce temporal changes in the death rate of all individuals in the community, while keeping total community size constant. With respect to changes in species relative abundance this still constitutes only demographic stochasticity as all species are equally affected by the prevailing death rate at a given time and each individual in the community has the same probability of survival per unit time. Neutral community models lack specific environmental stochasticity that would change the relative abundance of species, which in population genetic terms is analogous to fluctuating selection that randomly favors different alleles at different times (Gillespie, 1991). Null models of community structure (Gotelli and Graves, 1996) make fewer or different assumptions than neutral models but are still fundamentally based on ecologically equivalent species.

An important implication of the assumptions in neutral community models concerns the expected lifetime of common species before they become extinct. The lack of specific environmental stochasticity entails that in communities with a consistently large total size, N, the frequency of common species in the community will change very slowly, because demographic stochasticity exerts a substantial effect only in small populations (Chapters 1 and 2). By analogy to a neutral genetic model (Wright, 1931; Kimura, 1983), in a neutral community the expected rate of loss of Simpson diversity, $1 - \Lambda$ (eq. 7.5a), by demographic stochasticity is $1/N$ per generation. Thus the time scale for extinction of common species in a neutral community is on the order of N generations. For extremely abundant communities, such as oceanic plankton, tropical insects, and even tropical trees, this predicts that extinction of common species is not expected to occur within the age of the earth, whereas species observed in the fossil record become extinct within a few to several million years (Table 2.1).

The only escape from this problem for neutral community models is to suppose that total community size undergoes repeated restrictions to a small fraction of the average size, or that there is (geologically) frequent local extinction and re-expansion, both of which would substantially reduce the effective size of the community in comparison with its average actual size (Wright, 1940; Kimura, 1983). (By analogy with a neutral genetic model, the effective size of a neutral community with changing size and geographic structure is the size of an ideal neutral community of constant size that would produce the same expected rate of loss of Simpson diversity (Wright, 1931; Kimura, 1983).) However, reducing the effective community size enough to shorten the expected longevity of common species to realistic values also would substantially reduce the expected species richness in the community (Ewens, 1969; Kimura, 1983). Therefore, it appears necessary to postulate that reducing the effective size of a neutral community also produces high rates of speciation. Repeated episodes of community reduction, fragmentation into multiple refugia, and re-expansion, have been discussed among possible mechanisms that may produce high rates of speciation in tropical communities (Huston, 1994; Brown and Lomolino, 1998; Schemske, 2002).

Even without the assumption of neutrality, we concluded in Chapter 2.6 that environmental stochasticity observed in typical ecological time series often predicts excessive species longevities by failing to account for nonstationary environments and evolution. Neutral and nonneutral community models alike, therefore, encounter some difficulty in accounting for long-term large-scale processes of speciation and extinction, unless we invoke nonstationary environmental changes and evolutionary mechanisms that are beyond the scope of purely ecological models (e.g. Pease *et al.*, 1989; Lande and Shannon, 1996; Schluter, 2000).

The stochastic community models on which the remainder of this chapter focuses are more general than neutral or null models. We are especially interested in analyzing complex communities composed of many taxonomically related and ecologically similar species, containing possibly more species than can stably coexist in the available niche space in a constant environment. The following models cannot answer the long-term questions of what determines species diversity or species longevity in a community, although they do address short-term factors that help to maintain species diversity and species abundance distributions. Our goal is to develop methods to estimate the amount of ecological heterogeneity among species in a community, and the extent to which species respond differently to environmental fluctuations, by analyzing a community sampled intensively over limited temporal and spatial scales in a spatially homogeneous habitat.

8.3 Lognormal species abundance distribution

In contrast to neutral community models that assume no species-specific environmental stochasticity and no ecological heterogeneity among species, we develop models that allow species abundances to change by specific environmental stochasticity in density-independent growth rate and deterministic density dependence within species. These species-specific factors, as well as general community-wide competition and environmental effects, cause fluctuations in the total community size. We assume that a community of taxonomically related species in the same trophic guild, such as insectivorous birds or mammalian carnivores, compete for the same or similar mutually limiting resources. Following this approach Engen and Lande (1996*a,b*) showed that different forms of intraspecific density dependence produce different shapes of species abundance distributions in the community. Logistic density dependence within species produces a Gamma distribution of species abundances. Gompertz density dependence within species (eq. 1.9) produces a lognormal species abundance distribution, which, under certain conditions, applies even with ecological heterogeneity among species. We now develop this lognormal model and extend it to analyze the spatio-temporal dynamics of a tropical butterfly community.

Denote the abundance of species i as N_i, and its log abundance as $X_i = \ln N_i$. The vector of log abundances in a community with S species is $\mathbf{X} = (X_1, X_2, \ldots, X_S)$. The dynamics of X_i are described by the stochastic differential equation (Chapter 2), including terms for intra- and interspecific density dependence, demographic stochasticity, and both general and specific environmental stochasticity

$$\frac{dX_i}{dt} = r_i + \frac{\sigma_d}{\sqrt{N_i}}\frac{dA_i(t)}{dt} + \sigma_e\frac{dB_i(t)}{dt} - \gamma X_i - \Omega(\mathbf{X}) + \sigma_E\frac{dE(t)}{dt}. \qquad 8.1a$$

The density-independent long-run growth rate of species i is r_i and fluctuations in the growth rate are produced by demographic variance σ_d^2 and species-specific environmental variance σ_e^2 (Chapter 1). Density dependence acts deterministically within and among species. Species-specific density dependence, with the same Gompertz form for all species in the community, γX_i, could be caused by various intraspecific competition and crowding mechanisms described above, including the Janzen–Connell effect. Competition for common resources such as space and nutrients exerts the same density limitation on the growth rate of all species, $\Omega(\mathbf{X})$, and fluctuations in the general community environment identically affect the growth rate of each species, with general environmental variance σ_E^2. Stochastic effects are described by white noise (with no temporal autocorrelation), given by the derivatives of the independent Brownian motions for the species-specific demographic and environmental effects $A_i(t)$, $B_i(t)$, and the general community environment $E(t)$.

Expressing equation 8.1a in incremental form,

$$dX_i = [r_i - \gamma X_i - \Omega(\mathbf{X})]\,dt + \frac{\sigma_d}{\sqrt{N_i}}\,dA_i(t) + \sigma_e\,dB_i(t) + \sigma_E\,dE(t) \qquad \text{8.1b}$$

the stochastic dynamics of species abundances are governed by a multivariate diffusion process with infinitesimal means, $M_i(X_i) \equiv (1/dt)\mathrm{E}[dX_i|X_i]$, and infinitesimal variances and covariances, $V_{ij}(X_i, X_j) \equiv (1/dt)\mathrm{Cov}[dX_i, dX_j|X_i, X_j]$, derived using the Ito stochastic calculus (Chapter 2.3),

$$M_i(X_i) = r_i - \gamma X_i - \Omega(\mathbf{X})$$

$$V_{ij}(X_i, X_j) = \delta_{ij}\left(\frac{\sigma_d^2}{N_i} + \sigma_e^2\right) + \sigma_E^2 \qquad \text{8.2}$$

where $\delta_{ij} = 1$ if $i = j$ and 0 otherwise.

Random samples from a community (without mark–recapture data) are more informative about relative than absolute abundance of species. Therefore we analyze the stochastic dynamics of the mean log abundance of species in the community, $\bar{X} = (1/S)\sum_{k=1}^{S} X_k$, and the relative log abundances defined as deviations from the mean log abundance in the community, $x_i = X_i - \bar{X}$ for $i = 1, \ldots, S$, which entails that the relative log abundances sum to 0.

First consider the dynamics of the mean log abundance. Taking increments in the definition of the mean log abundance, $d\bar{X} = (1/S)\sum_{k=1}^{S} dX_k$, calculating the expectation and variance of $d\bar{X}$, and using equations 8.2, produces the infinitesimal mean and infinitesimal variance

$$M(\bar{X}) = \bar{r} - \gamma \bar{X} - \Omega(\mathbf{X}) \quad \text{and} \quad V(\bar{X}) = \sigma_E^2 + \frac{1}{S}\left(\sigma_e^2 + \frac{\sigma_d^2}{\tilde{N}}\right) \qquad \text{8.3}$$

where $\bar{r} = (1/S)\sum_{k=1}^{S} r_k$ is the mean density-independent long-run growth rate in the community, and $\tilde{N}$ is the harmonic mean abundance $1/\tilde{N} = (1/S)\sum_{k=1}^{S}(1/N_k)$. Comparison of the infinitesimal mean in equation 8.3 with that in 8.4 indicates that perturbations from equilibrium in the mean log abundance should be damped at least as rapidly as those for the relative log abundances. The infinitesimal variance shows that in a community with high species richness, $S \gg 1$, almost all of the stochasticity in mean log abundance is caused by general environmental variance σ_E^2 and that specific environmental variance and demographic stochasticity are relatively unimportant.

We now turn to the dynamics of the relative log abundances. Taking increments in the definition of relative log abundance, $dx_i = (1 - 1/S)\,dX_i - (1/S)\sum_{k\neq i} dX_k$, calculating expectations, products, and cross-products of the dx_i, and using equations 8.2, produces the infinitesimal means and infinitesimal covariances

$$\mu_i(x_i) = r_i - \bar{r} - \gamma x_i$$

$$v_{ij}(x_i, x_j) = \left(\delta_{ij} - \frac{1}{S}\right)\sigma_e^2 + \left(\frac{\delta_{ij}}{N_i} - \frac{1}{SN_i} - \frac{1}{SN_j} + \frac{1}{S\tilde{N}}\right)\sigma_d^2. \qquad 8.4$$

The multivariate diffusion for relative log abundances has an especially simple form. The dynamics of relative log abundances do not depend on the common factors for interspecific competition, $\Omega(\mathbf{X})$, or general environmental variance, σ_E^2, which change the log abundances of all species by the same amount. The infinitesimal mean for each species depends only on its own relative log abundance. Intraspecific density dependence, γ, drives every species toward the same abundance ($\bar{x} = 0$), while ecological heterogeneity among species in density-independent long-run growth rate causes species with r_i greater (less) than $\bar{r}$ to increase (decrease) in relative log abundance. Together, these forces act to maintain a diversity of species in the community with consistent differences in log abundance, ranging from relatively common to relatively rare, despite stochastic factors that would eventually cause the extinction of every species.

To show this more explicitly, we derive the multivariate stationary distribution of relative log abundances in the community, neglecting demographic stochasticity. Assuming a very large total community size with only a small fraction of the community included in samples, species with small total numbers are unlikely to be sampled, which justifies neglecting demographic stochasticity. Furthermore, simple demographic models do not allow for adaptive evolution, which is likely to be a major factor in the long-term persistence of species in changing environments, especially with nonstationary changes such as long-term cycles and trends that also are absent from the present model (see Chapter 2.6.3; Lande and Shannon, 1996). This limits the validity of simple demographic models for investigating long-term processes of species turnover and maintenance of species diversity by speciation and extinction. Because detailed empirical studies of community dynamics are necessarily short term, in developing the theory it is appropriate to condition on the distribution of r_i among species in the community at a given time by using the stationary distribution (eq. 2.4) instead of the quasi-stationary distribution (after eq. 2.6a).

In the diffusion for relative log abundances (eqs 8.4) the infinitesimal means are linear functions and, neglecting demographic stochasticity, the infinitesimal variances and covariances are constants. This constitutes a multivariate Ornstein–Uhlenbeck process that, beginning from a given state of the community, has a dynamic solution that is multivariate normal. The solution is straightforward. Because the infinitesimal means are independent among species, the dynamics of the distribution of relative log abundance in each species can be obtained from the solution of a univariate diffusion process (analogous to eqs 2.2b,c), which is easily extended to include the covariances,

$$E[x_i(t)] = \frac{r_i - \bar{r}}{\gamma} + \left[x_i(0) - \frac{r_i - \bar{r}}{\gamma} \right] e^{-\gamma t} \qquad 8.5a$$

$$Cov[x_i(t), x_j(t)] = \left(\delta_{ij} - \frac{1}{S} \right) \frac{\sigma_e^2}{2\gamma} (1 - e^{-2\gamma t}). \qquad 8.5b$$

Thus in the stationary distribution (as $t \to \infty$) the relative log abundance of species i has mean $(r_i - \bar{r})/\gamma$ and variance $(1 - 1/S)\sigma_e^2/(2\gamma)$. At any time the correlation of relative log abundances between each pair of species, $-1/(S - 1)$, is negative because of the constraint that the relative log abundances sum to 0. In a community with many species, $S \gg 1$, the dynamics of relative log abundances are only weakly correlated among the species.

The species abundance distribution in the community is the probability density function of relative log abundance of all the species. For analyzing data on community dynamics we are especially interested in the form of the species abundance distribution and its expected variance at the stationary distribution of relative log abundances in the community. Let $x(t)$ represent the relative log abundance of any species in the community. Because the mean relative log abundance in the community is always 0, the expected variance of $x(t)$ in the community equals the expected mean squared relative log abundance. At the stationary distribution we can write $x_i(t) = (r_i - \bar{r})/\gamma + \varepsilon_i(t)$ where $\varepsilon_i(t)$ is the deviation of the relative log abundance of species i at time t from its mean value, such that $E[\varepsilon_i(t)] = 0$ and $E\{[\varepsilon_i(t)]^2\} = (1 - 1/S)\sigma_e^2/(2\gamma)$ (eq. 8.5b). Finally, we assume that the r_i represent a random sample from a hypothetical distribution of density-independent long-run growth rates with mean $\mu_r = E[\bar{r}]$ and variance $\sigma_r^2 = E[(S - 1)^{-1} \sum_{i=1}^{S} (r_i - \bar{r})^2]$ (Stuart and Ord, 1994). Putting these statements together produces the expected variance of relative log abundances in the community at the stationary distribution,

$$E[Var[x(t)]] = E \left\{ \frac{1}{S} \sum_{i=1}^{S} [x_i(t)]^2 \right\}$$

$$= E \left\{ \frac{1}{S} \sum_{i=1}^{S} \frac{(r_i - \bar{r})^2}{\gamma^2} \right\} + E \left\{ \frac{1}{S} \sum_{i=1}^{S} [\varepsilon_i(t)]^2 \right\}$$

$$= \left(1 - \frac{1}{S} \right) \left(\frac{\sigma_r^2}{\gamma^2} + \frac{\sigma_e^2}{2\gamma} \right). \qquad 8.6$$

The form of the species abundance distribution can be derived as follows. First, recall that the joint stationary distribution of relative log abundances in the community is multivariate normal, and that in a community with many species the relative log abundances are nearly independent among species (eqs 8.5). In this model the univariate (marginal) stationary distribution of relative log abundance for a single species has the same shape but a different mean for each species. Consider a hypothetical community containing an infinite number of species with independent distributions of relative log abundance. In the infinite community let the distribution of density-independent long-run growth rates, r, among the species have mean μ_r and variance σ_r^2. A species in the infinite community with density-independent long-run growth rate r has a stationary distribution of relative log abundance, x, that is normal

$$\phi(x - \bar{x}) = \sqrt{\frac{\gamma}{\pi \sigma_e^2}} \exp\left\{-\frac{\gamma}{\sigma_e^2}(x - \bar{x})^2\right\}$$

with mean $\bar{x} = (r - \mu_r)/\gamma$ and variance $\sigma_e^2/(2\gamma)$. The distribution of $\bar{x}$ among species, denoted as $F(\bar{x})$, then has mean 0 and variance σ_r^2/γ^2. The species abundance distribution, $\Phi(x)$, is a convolution of the distribution of relative log abundance within species, $\phi(x - \bar{x})$, and the distribution of mean relative log abundances among species, $F(\bar{x})$, so that

$$\Phi(x) = \int_{-\infty}^{\infty} \phi(x - \bar{x}) F(\bar{x}) \, d\bar{x}. \qquad 8.7$$

This distribution of relative log abundance among species has mean 0 and variance $\sigma_r^2/\gamma^2 + \sigma_e^2/(2\gamma)$ because the variance of a convolution equals the sum of the variances of the two component distributions (Stuart and Ord, 1994).

Random sampling of S species from the species abundance distribution $\Phi(x)$ produces the expected variance in equation 8.6. At any time the distribution of log abundances, X_i, among these S species has the same shape as the distribution of their relative log abundances, x_i, but their mean log abundance, $\bar{X}$, fluctuates through time according to equations 8.3.

Finally, we show in this model that assuming a normal distribution of density-independent long-run growth rates among species implies a lognormal species abundance distribution. Assuming in the infinite community that r is normally distributed entails that $\bar{x}$ also is normally distributed. This and normality of the stationary distribution of relative log abundance in each species, $\phi(x - \bar{x})$, imply that the distribution of relative log abundances in the infinite community, $\Phi(x)$, is normal because a convolution of normal distributions also is normal (Stuart and Ord, 1994). Population sizes, $N_i = \exp\{x_i + \bar{X}\}$, among species in the community are therefore expected to be lognormally distributed.

8.4 Analyzing community dynamics in space and time

The foregoing model of stochastic community dynamics, generating a lognormal species abundance distribution, can be extended to encompass spatial distribution as well as temporal fluctuations, by drawing on results from previous chapters, following Engen *et al.* (2002c). Rather than testing neutral models that lack specific environmental stochasticity and ecological heterogeneity among species, we develop a more general framework for analyzing data on the stochastic dynamics of a community sampled through time across a homogeneous habitat. We partition the variance in the logarithmic species abundance distribution into components due to ecological heterogeneity among species, specific and general environmental stochasticity, and spatial overdispersion within species (including demographic stochasticity and intraspecific aggregation). We also analyze the strength of intraspecific density dependence and the scales of spatial autocorrelation in specific and general environmental stochasticity.

8.4.1   Relative log abundances within samples

Fluctuations in the relative abundance of a species at different spatial locations will be correlated because of spatial autocorrelation in specific environmental stochasticity and individual dispersal (Chapter 4.4), which we assume to be the same for all species in the community. Let $x_i(z,t)$ denote the relative log abundance of species i in location z at time t. Considering two populations of species i separated by displacement vector y and time lag τ, the covariance of their relative log abundances (approximated by eq. 4.12b) can be expressed as $v_{xw}\rho_w(y, \tau)$ where v_{xw} is the variance of relative log abundance within species through time at any given site (cf. eq. 8.5b), and $\rho_w(y, \tau)$ is the intraspecific spatio-temporal autocorrelation or synchrony due to individual dispersal and spatially autocorrelated specific environmental stochasticity. In Chapter 4.4 we noted that regional environmental fluctuations often create a long-distance component of population synchrony within species. This is most likely caused by regional fluctuations in the general environment that also exert parallel effects on the population growth rate and log abundance of all species in the community, without changing their relative abundances, as suggested by spatio-temporal data on three species of tetronid birds in Finland (Lindström *et al.*, 1996). Intraspecific synchrony in relative log abundance may therefore be small or absent at long distances, $\rho_w(\infty, \tau) = 0$.

An additional component of covariance in the community is created by variance among species in the means of their relative log abundances, v_{xa} (see after eqs 8.6, 8.7). With respect to the grand mean log abundance of species in the community the spatio-temporal autocovariance of any species is then the sum of the two components of covariance within and among species,

$$C(y, \tau) \equiv \mathrm{Cov}[x(z, t), x(z + y, t + \tau)] = v_{xw}\rho_w(y, \tau) + v_{xa}. \qquad 8.8a$$

In practice, the variance in relative log abundance among species at any given site contains a third component, v_o, caused by overdispersion of local population density (including demographic stochasticity and intraspecific aggregation) assumed to occur independently among species (see Chapter 7, after eq. 7.2), so that

$$v_x \equiv \mathrm{Var}[x(z, t)] = v_{xw} + v_{xa} + v_o. \qquad 8.8b$$

The spatio-temporal autocorrelation of relative log abundance of species in the community, $\rho_x(y, \tau)$, is the ratio of these two expressions,

$$\rho_x(y, \tau) = P_w \rho_w(y, \tau) + P_a \qquad \qquad 8.8c$$

where $P_w = v_{xw}/v_x$ and $P_a = v_{xa}/v_x$. Because of overdispersion within species, the community autocorrelation does not approach unity in the limit of short distance at zero time lag, $\rho_x(0, 0) = 1 - v_o/v_x$, although by definition the intraspecific synchrony in relative log abundance does, $\rho_w(0, 0) = 1$. For any time lag, intraspecific synchrony in relative log abundance vanishes at long distance, $\rho_w(\infty, \tau) = 0$, but ecological heterogeneity among species causes the community autocorrelation to remain positive at long distance, $\rho_x(\infty, \tau) = v_{xa}/v_x$, reflecting permanent differences among species in their mean relative log abundances across the region.

An estimate of the community autocorrelation in relative log abundances for a pair of samples can be obtained by fitting the observed joint species abundance distributions in the two samples to a bivariate Poisson-lognormal distribution (Engen *et al.*, 2002c). Assuming that the variance of log species abundance is the same at all sampling sites and times, this can be estimated using the average of the estimated variances obtained by fitting univariate Poisson-lognormal distributions to the observed species abundance distribution in each sample. Then using formulas 8.8a–c it is possible to estimate the spatio-temporal autocorrelation of the community and the three components of the variance of relative log abundance among species within sites and times, excluding the variance from random (Poisson) sampling.

The exact form of spatio-temporal autocorrelation within species, $\rho_w(y, \tau)$, depends on the strength of intraspecific density dependence per year, γ, and the usually unknown forms of spatial environmental autocorrelation and individual dispersal. The approximate formula for small or moderate fluctuations (eq. 4.12b) reveals a dominant temporal factor $e^{-\gamma \tau}$, which also is the temporal autocorrelation function in the Gompertz model of intraspecific density dependence allowing large fluctuations (see after eq. 2.2c). The spatial scale of population autocorrelation increases with increasing time lag between samples because of individual dispersal during the time lag (see after eq. 4.12b). However, if density dependence is strong, this effect will be difficult to detect because for time lags of more than a few years the factor $e^{-\gamma \tau}$ will make $\rho_w(y, \tau)$ small and $\rho_x(y, \tau)$ will become nearly flat at all spatial scales. Furthermore, in samples from communities containing many species, the sample sizes for particular species usually will not be large at each sampling locality and time. Hence substantial uncertainty will exist in the correlation of community composition estimated from pairs of samples, making it difficult to accurately estimate the spatial dependence of the community autocorrelation. Nevertheless, with numerous samples in space and time, it should be possible to estimate the spatial and temporal scales of community autocorrelation, the three components of variance in relative log abundance within samples (eq. 8.8b), and confidence intervals for these parameters (see below).

8.4.2 Mean log abundances among samples

General environmental effects do not alter the relative log abundance of species at any given site and time, but change their absolute log abundances in parallel. The impact of general environment on the community at a particular location, z, at time t can be estimated from the mean log abundance of species within a sample, $\bar{X}(z, t)$, obtained by fitting a univariate Poisson-lognormal distribution to the observed species abundance distribution in the sample. The variance and spatio-temporal autocorrelation in mean log abundance can be estimated using the method of maximum likelihood, assuming that the estimates $\bar{X}(z, t)$ come from a multivariate normal distribution (of dimension given by the number of samples) with equal variances, $v_{\bar{X}}$, and spatio-temporal autocorrelation function $\rho_{\bar{X}}(y, \tau)$ between distinct samples separated by displacement y and time lag τ. Components of $v_{\bar{X}}$ can be defined and estimated by assuming that the spatio-temporal autocorrelation can be approximated as a product of a temporal autocorrelation function, $\rho_{\bar{X}_\tau}(\tau)$, and a spatial autocorrelation function with a long-distance component,

$$\rho_{\bar{X}}(y, \tau) = \rho_{\bar{X}_\tau}(\tau)[Q_L q(\mathsf{R}) + Q_R] \qquad 8.8d$$

where $\mathsf{R} = \sqrt{y_1^2 + y_2^2}$ is the distance between samples in the local autocorrelation function, $q(\mathsf{R})$, with $q(0) = 1$ and $q(\infty) = 0$. Defining $Q_L = v_{\bar{X}L}/v_{\bar{X}}$ and $Q_R = v_{\bar{X}R}/v_{\bar{X}}$, the variance due to general environment can be partitioned into a sum of components as $v_{\bar{X}} = v_{\bar{X}L} + v_{\bar{X}R} + v_{\bar{X}\varepsilon}$ of which $v_{\bar{X}L}$ is involved in local autocorrelation, $v_{\bar{X}R}$ is due to long-distance regional effects, and $v_{\bar{X}\varepsilon}$ is a random component with no spatial autocorrelation. From estimates of $v_{\bar{X}}$ and the coefficients in the spatio-temporal autocorrelation function 8.8d, including Q_L and Q_R, these formulas permit estimation of the three components of $v_{\bar{X}}$.

8.5 Dynamics of a tropical butterfly community

We analyzed the stochastic dynamics of a fruit-feeding nymphalid butterfly community in mature rainforest at a field site in lowland Amazonian Ecuador (DeVries and Walla, 2001). Butterflies were sampled using fermenting banana bait at paired traps suspended in canopy and understory for 5 days per month, every month for 5 years at 25 locations within an approximately $2\,\text{km}^2$ area. The total sample comprised 11,861 individuals from 128 species.

8.5.1 Partitioning the variance in log abundance

Canopy and understory samples at each locality were pooled across the months within each year to avoid seasonal variation and to increase sample sizes. With 125 community samples categorized by locality and year, the average sample size was 95 individuals; the consequent large uncertainty in community correlation between samples, and the strong density dependence estimated below, make it difficult to estimate accurately the detailed form of the community autocorrelation function. For these reasons, and because we are primarily interested in the spatial and temporal scales of the community autocorrelation rather than its detailed form, we employed in equation 8.8c a simple exponential function for the spatio-temporal autocorrelation within species, $\rho_w(y, \tau) \equiv e^{-\gamma\tau - R/l_w}$, where R is the distance corresponding to displacement vector y. This allows us to estimate the temporal and spatial scales of intraspecific synchrony in relative log abundance as $1/\gamma$ and l_w, respectively, by fitting data on relative log abundance of species within community samples in space and time to the formula

$$\rho_x(y, \tau) = P_w e^{-\gamma\tau - R/l_w} + P_a. \qquad 8.8e$$

Alternatively, if we had assumed that intraspecific synchrony in relative log abundance has a positive long-distance component at zero time lag, ρ_∞, writing $\rho_w(y, \tau) = e^{-\gamma\tau}[(1 - \rho_\infty)e^{-R/l_w} + \rho_\infty]$, then equation 8.8c at $\tau = 0$ takes the same form as 8.8e but with P_w reduced by $\rho_\infty v_{xw}/v_x$ and P_a increased by the same amount. In estimating components of v_x from the community autocorrelation at $\tau = 0$, this would transfer $\rho_\infty v_{xw}$ from the variance caused by specific environmental stochasticity to that due to ecological heterogeneity, without changing the total v_x. The transferred variance is manifested at any given time as regional differences in mean relative log abundance among species. It therefore cannot be distinguished from permanent ecological heterogeneity among species, except by fitting separate spatial autocorrelation functions for the community at different time lags, which is not feasible with the present data because of strong density dependence (see below).

Observed abundances in pairs of distinct samples were fit to a bivariate Poisson-lognormal distribution (Engen *et al.*, 2002c) to estimate the community correlation for each of 7,750 pairs of distinct samples. Each estimated community correlation for a pair of samples has a large uncertainty that depends in an unknown way on the sample sizes and the actual community correlation. We therefore applied a simple least squares method to estimate the four constants in equation 8.8c from the set of 7,750 estimated community correlations, as illustrated in Figure 8.1. As our model assumes a constant variance of the normal distribution of log abundance of species at any location and time, we estimated v_x in equation 8.8b from the average value of the estimated variances in all pairs of samples. From equations 8.8a–c we then estimated the three components of v_x as well as the temporal and spatial scales of community autocorrelation. The 95% confidence intervals (in parentheses) for these parameters are based on 1,000 bootstrap values each obtained by resampling with replacement from the 7,750 community correlations to re-estimate the community autocorrelation function and v_x (Cressie, 1993; Bjørnstad *et al.*, 1999b).

Similarly, for general environmental effects we are most interested in the temporal and spatial scales of autocorrelation, as well as the component of variance among samples in mean log abundance, $v_{\bar{X}}$. We therefore chose exponential forms for the

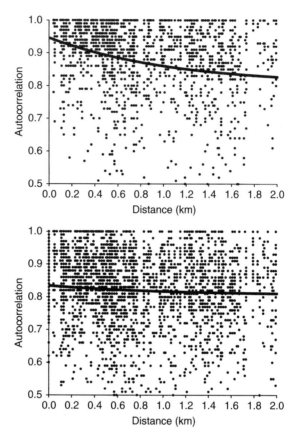

Fig. 8.1 Estimated community correlations for relative log abundance of species plotted against spatial distance between samples for time lag $\tau = 0$ (upper panel) and $\tau = 1$ yr (lower panel). The correlation at each point was estimated by fitting a bivariate Poisson-lognormal distribution to the observed species abundances in a pair of samples. The curves represent the fitted autocorrelation function given by equation 8.8e for $\tau = 0$ or 1, respectively. (From Engen *et al.*, 2002.)

temporal autocorrelation function, $\rho_{\bar{X}\tau}(\tau) = e^{-\alpha\tau}$ and the local spatial autocorrelation function $q(\mathsf{R}) = e^{-\mathsf{R}/l_{\bar{X}}}$ in equation 8.8d. This allows us to estimate the temporal and spatial scales for general environmental effects as $1/\alpha$ and $l_{\bar{X}}$, respectively, by fitting data on the mean log abundance of species in community samples in space and time to the formula

$$\rho_\mu(y, \tau) = e^{-\alpha\tau}[Q_L e^{-\mathsf{R}/l_{\bar{X}}} + Q_R]. \qquad 8.8f$$

From the infinitesimal mean in equations 8.3, we anticipate that α should be at least as large as the intraspecific density dependence, γ. The fitted spatio-temporal auto-correlation function for variation due to general environmental effects is depicted in

Figure 8.2. The 95% confidence intervals (in parentheses) for estimates of $v_{\bar{X}}$ and its components were obtained by parametric bootstrapping, simulating 1,000 data sets from the model using the estimated variance and autocorrelation function (Efron and Tibshirani, 1993).

The estimated variance of relative log abundance among species within samples, excluding variance from random (Poisson) sampling, was $\hat{v}_x = 3.418$, with estimated components shown in Table 8.1. The great majority of this is due to ecological

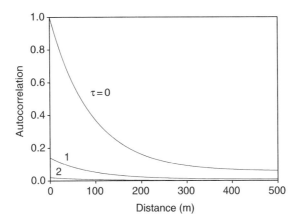

Fig. 8.2 Estimated spatio-temporal autocorrelation function for variation in mean log abundance due to general environment as a function of distance between samples for time lag τ years. The mean log abundance of species at each locality and time was estimated by fitting a Poisson-lognormal distribution to the observed species abundances in each sample.

Table 8.1 Estimated components of variance in log abundance of species within and among samples in a tropical butterfly community, excluding variance from random (Poisson) sampling within sampling units

Component of variance	Estimate	95% confidence interval
Within samples, v_x	3.418	(3.385, 3.454)
Ecological heterogeneity among species, v_{xa}	2.754	(2.706, 2.790)
Specific environment, v_{xw}	0.480	(0.393, 0.575)
Overdispersion (including demographic stochasticity), v_o	0.183	(0.106, 0.269)
Among samples (general environment), $v_{\bar{X}}$	1.237	(0.882, 1.445)
Locally correlated, $v_{\bar{X}L}$	1.166	(0.274, 1.398)
Regionally correlated, $v_{\bar{X}R}$	0.071	(0.000, 0.096)
Random (uncorrelated), $v_{\bar{X}\varepsilon}$	0.000	(0.000, 0.848)

heterogeneity among species, v_{xa}. Interspecific heterogeneity in density-independent growth rates could be caused by differences in reproduction and/or mortality. Differential reproduction likely arises from unequal abundances of acceptable larval host plant species (DeVries, 1987), and differential mortality may arise from unequal tolerance to environmental factors. A substantial component of variance in relative abundance among species within samples is attributable to specific environmental variation, v_{xw}, and a relatively small component to overdispersion (including demographic stochasticity), v_o.

The strength of intraspecific density dependence per year was estimated as $\hat{\gamma} = 1.61 \, \text{yr}^{-1}$ (0.58, 2.63), so the temporal scale of autocorrelation within species and in community composition, $1/\gamma$, is estimated to be less than 1 year. The strength of intraspecific density dependence per year in this butterfly community is large relative to that for long-lived vertebrates, as indicated by estimates of γ in Table 3.2, because the generation time is on the order of weeks or months for butterflies versus years or decades for long-lived vertebrates. High uncertainty in the estimate occurs not only because of modest sample sizes in each sampling unit, but also because the damping factor $e^{-\gamma\tau}$ decays to small values for time lags greater than 1 year, so that most pairs of samples contain little information on this parameter.

The estimated spatial scale of intraspecific synchrony in relative log abundance depends mainly on pairs of samples with time lag of 0 or 1 year, because of the strong intraspecific density dependence per year, as illustrated in Figure 8.1. The spatial scale of intraspecific synchrony in relative log abundance was estimated as $\hat{l}_w = 1.04 \, \text{km}$ (0.67, 1.61). The spatial autocorrelation of community composition remains high at long distances for any time lag due to ecological heterogeneity among species, $\rho_x(\infty, \tau) = 0.806$ (eq. 8.8c, Fig. 8.1).

The variance among samples in mean log abundance of the local community, caused by the general environment, was estimated as $\hat{v}_{\bar{X}} = 1.237$. This is smaller than the estimate 1.527 obtained using a different method by Engen et al. (2002c) who did not account for spatio-temporal autocorrelation of variation due to general environment, nor partition it into components as shown in Table 8.1. Nearly all of the variance due to general environment is locally autocorrelated. The spatial scale of autocorrelation in general environmental stochasticity was less than that for specific environmental stochasticity. The rate of return to the average μ was estimated as $\hat{\alpha} = 1.96 \, \text{yr}^{-1}$. As anticipated from equation 8.3, within the statistical accuracy of the estimates $\hat{\alpha}$ is greater than or equal to $\hat{\gamma}$. General environmental perturbations tend to be damped within about $1/\alpha = 0.510 \, \text{yr}$ (0.000, 0.759). The spatial scale of autocorrelation in general environmental stochasticity was estimated as $\hat{l}_{\bar{X}} = 0.091 \, \text{km}$ (0.000, 0.147). These results indicate that general environmental stochasticity in this fruit-feeding butterfly community tends to occur as small scale events that are rapidly damped (Fig. 8.2). General environmental stochasticity may arise from random annual variation in the fruit crop of individual canopy trees and local understory plants, although fruiting of tropical plants often displays seasonal synchrony within and among species (Foster, 1996; Leigh, 1999, ch. 7), annual variation in local fruit crops may nevertheless have a substantial random local component affecting the community of fruit-feeding butterflies.

8.5.2 Temporal fluctuations in Shannon information

Temporal changes in species diversity at a given site, measured by Shannon information H (eq. 7.4a), were modeled using components of variance in the logarithmic species abundance distribution in the community caused by ecological heterogeneity among species and specific environmental stochasticity (Table 8.1 and eq. 8.8e). To estimate the total number of species in the community we fit a univariate Poisson-lognormal distribution to the observed total abundance of each species in the pooled data set, giving $S = 158$. Temporal fluctuations in 158 species with an expected normal distribution of relative log abundances at a locality were simulated using equations 8.4, ignoring demographic stochasticity (overdispersion) and random (Poisson) sampling within sampling units. General environmental effects also were excluded as they do not alter species relative abundances within sites and therefore do not affect species diversity. At annual intervals, the simulated relative log abundances were converted to species frequencies in the community to compute H. Figure 8.3 shows that despite the large component of permanent differences in relative log abundance due to ecological heterogeneity among species, specific environmental stochasticity creates substantial annual fluctuations in H at a given site.

8.5.3 Implications of the results

The large component of variance due to ecological heterogeneity among species strongly violates the central assumption of neutral models of community structure

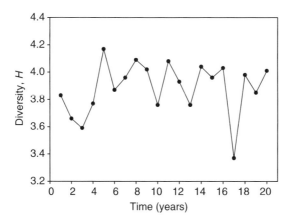

Fig. 8.3 Simulated temporal fluctuations in the Shannon information measure of species diversity, H, at a locality caused by specific environmental stochasticity, not including seasonal environmental effects, overdispersion among sampling units, or random (Poisson) sampling within sampling units. The simulation is based on the estimated variance in relative log abundance due to specific environmental stochasticity and ecological heterogeneity among species (Table 8.1) and the estimated total number of species in the community, $S = 158$, obtained by fitting a univariate Poisson-lognormal distribution to the observed species abundances in the entire data set of 11,861 individuals. (Modified from Engen *et al.*, 2002.)

based on ecologically interchangeable species. Neutral models imply that all species have the same (quasi-stationary) distribution of relative abundance, so that the relative abundance of particular species is purely a matter of chance. The assumption in neutral models of no ecological heterogeneity among species leads to the prediction that on spatial and temporal scales larger than those of intraspecific synchrony in relative log abundance there is no predictability to species relative abundances in space. In contrast, the large component of variance due to ecological heterogeneity among species (Table 8.1) implies that the relative abundance of species tends to be preserved across space and time as reflected in the high spatial autocorrelation of the community at long distances for any time lag (Fig. 8.1). This entails that rare species remain consistently rare, and that common species persist at high relative abundances, across long distances and through time.

Neutral models of stochastic community dynamics include only demographic stochasticity and omit specific environmental stochasticity. Table 8.1 also demonstrates that the components of variance in relative log abundance among species due to specific and general environmental stochasticity are much larger than that due to overdispersion (including demographic stochasticity). Thus the stochastic mechanism in neutral models explains only a small fraction of the total variance in local species abundance distributions for the butterfly community we studied.

These results indicate that the butterfly community we analyzed is far from neutral. If this conclusion can be extended to other communities, neutral models of community dynamics would lose their appeal even as null hypotheses, as they would be generally rejected. Neutral models could then be replaced by more general dynamic models permitting estimation of ecological heterogeneity among species, intraspecific density dependence, and specific and general environmental stochasticity.

From this analysis of a tropical butterfly community we learned that there is great consistency in species relative abundances through space and time. Nevertheless, specific environmental stochasticity produces substantial fluctuations in local species diversity through time, as measured by temporal fluctuations in Shannon information at one site (Fig. 8.3). We also learned that specific and general environmental perturbations tend to occur on small spatial scales and are rapidly damped by strong intraspecific density dependence. These results would be especially useful if we were to return to the same community after severe habitat degradation or other human impacts, and ask how the community dynamics had changed. Quantitative analysis of community dynamics should become an essential complement to population viability analyses of particular species. Of course, the present approach still represents a rather simplified description of a highly diverse and complex community. To reduce the number of model parameters to a tractable level for estimation from the data, we necessarily made simplifying assumptions, including spatial homogeneity of the habitat, symmetry among the species in terms of equal magnitude of specific environmental affects, the same functional form of intraspecific density dependence, interspecific interaction occurring through a single general competition function, and identical spatial scales of intraspecific synchrony in relative log abundance. Future work that focuses on relaxing these assumptions will be worthwhile, but the utility of models for analyzing the dynamics of complex communities always will be limited by the number of parameters that can be estimated with reasonable statistical confidence from the data.

8.6 Summary

In simple ecosystems with a few strongly interacting species, realistic models can be constructed to describe interspecific ecological interactions and the effects of environmental variables on the species, but such models usually are not feasible for complex ecosystems because of insufficient data. Ecologists have long been concerned with understanding how species diversity is maintained, especially in complex ecosystems containing communities of many taxonomically related species in the same trophic guild, such as insectivorous birds or mammalian carnivores. Deterministic theories of competitive exclusion entail that multiple species cannot coexist at a stable equilibrium on the same resources. Recent nonequilibrium theories have revealed various mechanisms for maintaining ecologically similar or identical species in a community. Ecologically similar species can coexist indefinitely on the same resources because of species-specific environmental stochasticity. Neutral community models maintain a diversity of ecologically equivalent species in a constant environment through a balance between speciation and extinction by demographic stochasticity.

We constructed stochastic models of a community of species in the same trophic guild, competing for the same or similar mutually limiting resources, and subject to species-specific density regulation as well as specific and general environmental stochasticity. Assuming intraspecific density dependence of the Gompertz form, and allowing ecological heterogeneity among species in their mean relative abundance, produces a lognormal species abundance distribution. We extended this model to describe spatial as well as temporal dynamics and applied it to analyze a community of fruit-feeding tropical butterflies sampled extensively in space and time. We fit a bivariate Poisson-lognormal species abundance distribution to each pair of community samples in space and time to estimate patterns of spatio-temporal covariance in community composition. The results demonstrate that the majority of variance in log abundance of species in the community is caused by ecological heterogeneity among species, implying a high consistency in their relative abundances across space and time. Specific environmental stochasticity nevertheless produces substantial temporal fluctuations in species diversity at a given site. Specific and general environmental stochasticity occur on short spatial scales and are rapidly damped by strong intraspecific density dependence. Only a small fraction of the variance in log abundance of species is caused by overdispersion (including demographic stochasticity) within species. This indicates that real communities are far from neutral.

References

Aanes, S., Engen, S., Sæther, B.-E., Willebrand, T., and Marcström, V. 2002. Sustainable harvesting strategies of Willow Ptarmigan in a fluctuating environment. *Ecological Applications*, **12**, 281–90.

Abrams, P. A. 1976. Niche overlap and environmental variability. *Mathematical Biosciences*, **28**, 357–75.

Abrams, P. A. 1984. Variability in resource consumption rates and the coexistence of competing species. *Theoretical Population Biology*, **25**, 106–24.

Aebischer, N. J. and Potts, G. R. 1990. Sample size and area: implications based on long-term monitoring on partridges. In *Pesticide effects on terrestrial wildlife* (eds L. Somerville and C. H. Walker), pp. 257–70. Taylor & Francis, London.

Aczél, J. and Daróczy, Z. 1975. *On measures of information and their characterizations.* Academic Press, New York.

Alatalo, R. and Alatalo, R. 1977. Components of diversity: multivariate analysis with interaction. *Ecology*, **58**, 900–6.

Allan, J. D. 1975a. Components of diversity. *Oecologia*, **18**, 359–67.

Allan, J. D. 1975b. The distributional ecology and diversity of benthic insects in Cement Creek, Colorado. *Ecology*, **56**, 1040–53.

Allee, W. C., Emerson, A. E., Park, O., Park, T., and Schmidt, K. P. 1949. *Principles of animal ecology.* W. B. Saunders Co., Philadelphia.

Allen, J. C., Schaffer, W. M., and Rostro, D. 1993. Chaos reduces species extinction by amplifying local population noise. *Nature*, **364**, 229–32.

Alroy, J. 2000. New methods for quantifying macroevolutionary patterns and processes. *Paleobiology*, **26**, 707–33.

Alroy, J. 2001. A multispecies overkill simulation of the end-Pleistocene megafaunal mass extinction. *Science*, **292**, 1893–6.

Alvarez-Buylla, E. and Slatkin, M. 1993. Finding confidence limits on population growth rates: Monte Carlo test of a simple analytical method. *Oikos*, **68**, 273–82.

Alverson, D. L., Freeberg, M. H., Pope, J. G., and Murawski, S. A. 1994. A global assessment of fisheries bycatch and discards. *FAO Fisheries Technical Paper*, **339**, 1–233.

Andrewartha, H. G. and Birch, L. C. 1954. *The distribution and abundance of animals.* University of Chicago Press, Chicago.

Ariño, A. and Pimm, S. L. 1995. On the nature of population extremes. *Evolutionary Ecology*, **9**, 429–43.

Armbruster, P. and Lande, R. 1993. A population viability analysis for African elephant (*Loxodonta africana*): How big should reserves be? *Conservation Biology*, **7**, 602–10.

Armbruster, P., Fernando, P., and Lande, R. 1999. Time frames for population viability analysis of species with long generations: an example with Asian elephants. *Animal Conservation*, **2**, 69–73.

Armstrong, R. A. and McGehee, R. 1980. Competitive exclusion. *American Naturalist*, **115**, 151–70.

Armsworth, P. R. and Roughgarden, J. E. 2001. An invitation to ecological economics. *Trends in Ecology & Evolution*, **16**, 229–34.

Azuma, S., Sasaki, T., and Itô, Y. 1997. Effects of undergrowth removal on the species diversity of insects in natural forests of Okinawa Hontô. *Pacific Conservation Biology*, **3**, 156–60.

Bacon, P. J. and Andersen-Harild, P. 1989. Mute Swan. In *Lifetime reproduction in birds* (ed. I. Newton), pp. 363–86. Academic Press, New York.

Bacon, P. J. and Perrins, C. M. 1991. Long term population studies: the mute swan. *Acta Congressus Internationalis Ornithologici*, **20**, 1500–13.

Ballard, W. G., Whitman, J. S., and Reed, D. J. 1991. Population dynamics of moose in south-central Alaska. *Wildlife Monographs*, **114**, 1–49.

Bartlett, M. S. 1960. *Stochastic population models in ecology and epidemiology*. Methuen, London.

Bascompte, J. and Solé, J. 1996. Habitat fragmentation and extinction thresholds in spatially explicit models. *Journal of Animal Ecology*, **65**, 465–73.

Beattie, A. J. (ed.). 1993. Rapid biodiversity assessment. *Proceedings of the biodiversity workshop, Macquarie University, Sydney*. Research Unit for Biodiversity and Bioresources, Macquarie University.

Beddington, J. R. and May, R. M. 1977. Harvesting natural populations in a randomly fluctuating environment. *Science*, **197**, 463–65.

Begon, M., Harper, J. H., and Townsend, C. R. 1996. *Ecology: individuals, populations and communities*. 3rd edn, Blackwell, Oxford.

Beier, P. 1993. Determining minimum habitat areas and habitat corridors for cougars. *Conservation Biology*, **7**, 94–108.

Beissinger, S. R. and McCullough, D. R. (eds). 2002. *Population viability analysis*. University of Chicago Press, Chicago.

Beissinger, S. R. and Westphal, M. I. 1998. On the use of demographic models of population viability in endangered species management. *Journal of Wildlife Management*, **62**, 821–41.

Bell, G. 2000. The distribution of abundance in neutral communities. *American Naturalist*, **155**, 606–17.

Bell, G. 2001. Neutral macroecology. *Science*, **293**, 2413–18.

Benton, T. G., Grant, A., and Clutton-Brock, T. H. 1995. Does environmental stochasticity matter? Analysis of red deer life-histories on Rum. *Evolutionary Ecology*, **9**, 559–74.

Berryman, A. A. and Turchin, P. 1997. Detection of delayed density dependence: comment. *Ecology*, **78**, 318–20.

Birkhead, M. R. and Perrins, C. 1986. *The Mute Swan*. Croom Helm, London.

Bjørnstad, O. N., Stenseth, N. C., and Saitoh, T. 1999a. Synchrony and scaling in dynamics of voles and mice in northern Japan. *Ecology*, **80**, 622–37.

Bjørnstad, O. N., Ims, R. A., and Xavier, L. 1999b. Spatial population dynamics: analyzing patterns and processes of population synchrony. *Trends in Ecology & Evolution*, **14**, 427–32.

Bjørnstad, O. N. and Grenfell, B. T. 2001. Noisy clockwork: Time series analysis of population fluctuations in animals. *Science*, **293**, 638–43.

Blums, P., Bauga, I., Leja, P., and Mednis, A. 1993. Breeding populations of ducks on Engure Lake, Latvia, for 35 years. *Ring*, **15**, 165–9.

Blums, P., Mednis, A., Bauga, I., Nichols, J. D., and Hines, J. E. 1996. Age-specific survival and philopatry in three species of European ducks: a long-term study. *Condor*, **98**, 61–74.

Bolker, B. M. and Grenfell, B. T. 1996. Impact of vaccination on the spatial correlation and the persistence of measels dynamics. *Proceedings of the National Academy of Sciences, USA*, **93**, 12648–53.

Botsford, L. W., Hastings, A., and Gaines, S. D. 2001. Dependence of sustainability on the configuration of marine reserves and larval dispersal distance. *Ecology Letters*, **4**, 144–50.

Bowman, K. O., Hutcheson, K., Odum, E. P., and Shenton, L. R. 1971. Comments on the distribution of indices of diversity. In *Statistical ecology*. Vol. 3. *Many species populations, ecosystems, and systems analysis* (eds G. P. Patil, E. C. Pielou, and W. E. Waters), pp. 315–66. Pennsylvania State University Press, University Park, Pennsylvania.

Boyce, M. S. 1992. Population viability analysis. *Annual Review of Ecology and Systematics*, **23**, 481–506.

Box, G. E. P., Jenkins, G. M., and Reinsel, G. C. 1994. *Time series analysis: forecasting and control*. 3rd edn, Prentice Hall, Englewood Cliffs, New Jersey.

Bradley, J. S., Skira, I. J., and Wooller, R. D. 1991. A long-term study of Short-tailed Shearwaters *Puffinus tenuirostris* on Fisher island, Australia. *Ibis*, **133** Supplement 1, 55–61.

Bro, E., Sarrazin, F., Clobert, J., and Reitz, F. 2000. Demography and the decline of the grey partridge *Perdix perdix* in France. *Journal of Applied Ecology*, **37**, 432–48.

Brook, B. W., O'Grady, J. J., Chapman, A. P., Burgman, M. A., Akçakaya, H. R., and Frankham, R. 2000. Predictive accuracy of population viability analysis in conservation biology. *Nature*, **404**, 385–7.

Brown, J. L. 1975. *The evolution of behavior*. Norton, New York.

Brown, J. H. and Kodric-Brown, A. 1977. Turnover rates in insular biogeography: effect of immigration on extinction. *Ecology*, **58**, 445–9.

Brown, J. H. and Lomolino, M. V. 1998. *Biogeography*. 2nd edn, Sinauer, Sunderland, Massachusetts.

Bryant, D., Burke, L., McManus, J., and Spalding, M. 1998. *Reefs at risk. A map-based indicator of threats to the world's coral reefs*. World Resources Institute, Washington, D.C.

Buckland, S. T., Anderson, D. R., Burnham, K. P., and Laake, J. L. 1993. *Distance sampling*. Chapman & Hall, New York.

Bull, J. J. 1983. *Evolution of sex determining mechanisms*. Benjamin Cummings, Menlo Park, California.

Bulmer, M. G. 1974. On fitting the Poisson lognormal distribution to species-abundance data. *Biometrics*, **30**, 101–10.

Bulmer, M. G. 1975. The statistical analysis of density dependence. *Biometrics*, **31**, 901–11.

Bunge, J. and Fitzpatrick, M. 1993. Estimating the number of species. A review. *Journal of the American Statistical Association*, **88**, 364–73.

Burgman, M. A., Ferson, S., and Akçakaya, H. R. 1993. *Risk assessment in conservation biology*. Chapman & Hall, New York.

Burnham, K. P. and Anderson, D. R. 1998. *Model selection and inference: a practical information—theoretic approach*. Springer-Verlag, New York.

Carey, J. R. 1993. *Applied demography for biologists, with special emphasis on insects*. Oxford University Press, New York.

Carroll, R. L. 1988. *Vertebrate paleontology and evolution*. W. H. Freeman, New York.

Case, T. J. 2000. *An illustrated guide to theoretical ecology*. Oxford University Press, New York.

Case, T. J., Bolger, D. T., and Richman, A. D. 1998. Reptilian extinctions over the last ten thousand years. In *Conservation biology for the coming decade*. 2nd edn (eds P. L. Fiedler and P. M. Kareiva), pp. 157–86. Chapman & Hall, New York.

Casey, J. M. and Myers, R. A. 1998. Near extinction of a large, widely distributed fish. *Science*, **281**, 690–2.

Caswell, H. 1976. Community structure: a neutral model analysis. *Ecological Monographs*, **46**, 327–54.

Caswell, H. 1978. A general formula for the sensitivity of population growth rate to changes in life history parameters. *Theoretical Population Biology*, **14**, 215–30.

Caswell, H. 2001. *Matrix population models*. 2nd edn, Sinauer, Sunderland, Massachusetts.

Caughley, G. 1977. *Analysis of vertebrate populations*. Wiley, London.

Caughley, G. 1987. Ecological relationships. In *Kangaroos: their ecology and management in the sheep rangelands of Australia* (eds G. Caughley, N. Shepherd and J. Short), pp. 159–87. Cambridge University Press, Cambridge.

Caughley, G. 1994. Directions in conservation biology. *Journal of Animal Ecology*, **63**, 215–44.

Caughley, G. and Gunn, A. E. 1993. Dynamics of large herbivores in deserts: kangaroos and caribou. *Oikos*, **67**, 47–55.

Caughley, G. and Gunn, A. E. 1996. *Conservation biology in theory and practice*. Blackwell Science, Cambridge, Massachusetts.

Chao, A. and Lee, S.-M. 1992. Estimating the number of classes via sample coverage. *Journal of the American Statistical Association*, **87**, 210–17.

Charlesworth, B. 1994. *Evolution in age-structured populations*. 2nd edn, Cambridge University Press, Cambridge.

Charnov, E. L. 1993. *Life history invariants*. Oxford University Press, Oxford.

Chatfield, C. 1996. *The analysis of time series. An introduction*. 5th edn, Chapman & Hall/CRC, Boca Raton.

Chepko-Sade, B. D. and Halpin, Z. T. (ed.). 1987. *Mammalian dispersal patterns*. University of Chicago Press, Chicago.

Chesson, P. 2000. Mechanisms of maintenance of species diversity. *Annual Review of Ecology and Systematics*, **31**, 343–66.

Chesson, P. L. and Case, T. J. 1986. Overview. Nonequilibrium community theories: Chance, variability, history, and coexistence. In *Community ecology* (eds J. Diamond and T. J. Case), pp. 229–39. Harper & Row, New York.

Chesson, P. and Huntly, N. 1997. The roles of harsh and fluctuating conditions in the dynamics of ecological communities. *American Naturalist*, **150**, 519–53.

Chesson, P. L. and Warner, R. W. 1981. Environmental variability promotes coexistence in lottery competitive systems. *American Naturalist*, **117**, 923–43.

Claessen, D., de Roos, A., and Persson, L. 2000. Dwarf and giants: cannibalism and competition in size-structured populations. *American Naturalist*, **155**, 219–37.

Clark, C. W. 1973. The economics of overexploitation. *Science*, **181**, 630–4.

Clark, C. W. 1990. *Mathematical bioeconomics. The optimal management of renewable resources*. 2nd edn, Wiley, New York.

Clark, C. W. 1996. Marine reserves and the precautionary management of fisheries. *Ecological Applications*, **6**, 369–70.

Clark, C. W. and Kirkwood, G. P. 1986. On uncertain renewable resource stocks: optimal harvest policies and the value of stock surveys. *Journal of Environmental Economics and Management*, **3**, 235–44.

Clobert, J., Perrins, C. M., McCleery, R. H., and Gosler, A. G. 1988. Survival rate in the great tit *Paris major* in relation to sex, age, and immigration status. *Journal of Animal Ecology*, **57**, 287–306.

Clutton-Brock, T. H. 1988. *Reproductive success*. University of Chicago Press, Chicago.

Cohen, J. 1977. Ergodicity of age structure in populations with Markovian vital rates, III: Finite-state moments and growth rate; an illustration. *Advances in Applied Probability*, **9**, 462–75.

Cohen, J. E. 1979. Ergodic theorems in demography. *Bulletin of the American Mathematical Society*, **1**, 275–95.

Cohen, J. E. 1995. *How many people can the earth support?* W.W. Norton & Co., New York.

Colwell, R. K. and Coddington, J. A. 1994. Estimating terrestrial biodiversity by extrapolation. *Philosophical Transactions of the Royal Society of London B*, **345**, 101–18.

Connell, J. H. 1971. On the role of natural enemies in preventing competitive exclusion in some marine animals and in rain forest trees. In *Dynamics of populations* (eds P. J. den Boer and G. R. Gradwell), pp. 298–312. Centre for Agricultural Publishing and Documentation, Wageningen, The Netherlands.

Connell, J. H. 1978. Diversity in tropical rain forests and coral reefs. *Science*, **199**, 1302–10.

Conservation International. 2002*a*. Conservation Strategy: Biodiversity Hotspots. http://www.conservation.org/xp/CIWEB/strategies/hotspots/hotspots.xml

Conservation International. 2002*b*. Center for Applied Biodiversity Science. CABS Research. Rapid Assessment Program. http://www.biodiversityscience.org/xp/CABS/research/rap/aboutrap.xml

Convention on Biological Diversity. 1992. http://www.biodiv.org/convention/articles.asp

Coope, G. R. 1979. Late Cenozoic fossil Coleoptera: evolution, biogeography, and ecology. *Annual Review of Ecology and Systematics*, **10**, 247–67.

Coulson, T., Catchpole, E. A., Albon, S. D., Morgan, B. J. T., Pemberton, J. M., Clutton-Brock, T. H., Crawley, M. J., and Grenfell, B. J. 2001*a*. Age, sex, density, winter weather, and population crashes in Soay sheep. *Science*, **292**, 1528–31.

Coulson, T., Mace, G. M., Hudson, E., and Possingham, H. 2001*b*. The use and abuse of population viability analysis. *Trends in Ecology & Evolution*, **16**, 219–21.

Courchamp, F., Clutton-Brock, T. H., and Grenfell, B. 1999. Inverse density dependence and the Allee effect. *Trends in Ecology & Evolution*, **14**, 405–10.

Cox, D. R. and Miller, H. D. 1965. *The theory of stochastic processes*. Chapman & Hall, New York.

Cramp, S. 1972. One hundred and fifty years of Mute Swans on the Thames. *Wildfowl*, **23**, 119–24.

Cressie, N. A. C. 1993. *Statistics for spatial data*. Wiley, New York.

Crête, M., Rivest, L.-P., Jolicoeur, H., Brassard, J.-M., and Messier, F. 1986. Predicting and correcting helicopter counts of moose with observations made from fixed-wing aircraft in Southern Quebec. *Journal of Applied Ecology*, **23**, 751–67.

Crist, T. O., Veech, J. A., Gering, J. C., and Summerville, K. S. 2002. Partitioning species diversity across landscapes and regions: a hierarchical analysis of α, β, and γ diversity. (in preparation)

Crooks, J. and Soulé, M. E. 1996. Lag times in population explosions of invasive species: causes and implications. In *Invasive species and biodiversity management* (eds O. Sandlund, P. J. Schei, and Å. Viken), pp. 39–46. Kluwer, Dordrecht.

Croxall, J. P. and Rothery, P. 1991. Population regulation of seabirds: implications of their demography for conservation. In *Bird population studies. Relevance to conservation and management* (eds C. M. Perrins, J.-D. Lebreton, and G. J. M. Hirons), pp. 272–96. Oxford University Press, Oxford.

Croxall, J. P., Prince, P. A., Rothery, P., and Wood, A. G. 1997. Population changes in albatrosses at South Georgia. In *Albatross biology and conservation* (eds G. Robertson and R. Gales), pp. 69–83. Surrey Beattie, Chipping Norton.

Cumming, D. H. M., Du Toit, R. F., and Stuart, S. N. 1990. *African elephants and rhinos: status survey and conservation action plans.* IUCN, Gland, Switzerland.

Danchin, E., Cadiou, B., Monnat, J.-Y., and Estrella, R. R. 1991. Recruitment in long-lived birds: conceptual framework and behavioural mechanisms. *Acta Congressus Internationalis Ornithologici*, **20**, 1641–56.

Davies, N. B. 1978. Ecological questions about territorial behavior. In *Behavioural ecology: an evolutionary approach* (eds J. R. Krebs and N. B. Davies), pp. 317–50. Sinauer, Sunderland, Massachusetts.

Davis, G. E., Haaker, P. L., and Richards, D. V. 1998. The perilous condition of white abalone *Haliotis sorenseni*, Bartsch, 1940. *Journal of Shellfish Research*, **17**, 871–5.

Davis, S. A., Pech, R. P., and Catchpole, E. A. 2002. Populations in variable environments: the effect of variability in a species' primary resource. *Philosophical Transactions of the Royal Society B*, **357**, 1249–59.

de Kroon, H., Plaisier, A., van Groenendael, J., and Caswell, H. 1986. Elasticity: the relative contribution of demographic parameters to population growth rate. *Ecology*, **67**, 1427–31.

Deevey, E. S, Jr. 1947. Life tables for natural populations of animals. *Quarterly Review of Biology*, **22**, 283–314.

Demetrius, L. 1969. The sensitivity of population growth rate to perturbations in the life cycle components. *Mathematical Biosciences*, **4**, 129–36.

Dennis, B. 1989. Allee effects: population growth, critical density, and the chance of extinction. *Natural Resource Modeling*, **3**, 481–538.

Dennis, B. 1996. Discussion: should ecologists become Bayesians? *Ecological Applications*, **6**, 1095–103.

Dennis, B. 2002. Allee effects in stochastic populations. *Oikos*, **96**, 389–401.

Dennis, B. and Otten, M. R. 2000. Joint effects of density dependence and rainfall on abundance of San Joaquin kit fox. *Journal of Wildlife Management*, **64**, 388–400.

Dennis, B. and Taper, M. L. 1994. Density dependence in time series observations of natural populations: estimation and testing. *Ecological Monographs*, **64**, 205–24.

Dennis, B., Munholland, P. L., and Scott, J. M. 1991. Estimation of growth and extinction parameters for endangered species. *Ecological Monographs*, **61**, 115–43.

Dennis, B., Desharnais, R. A., Cushing, J. M., Henson, S. M., and Costantino, R. F. 2001. Estimating chaos and complex dynamics in an insect population. *Ecological Monographs*, **71**, 277–303.

DeVries, P. J. 1987. *The butterflies of Costa Rica and their natural history. I: Papilionidae, Pieridae and Nymphalidae.* Princeton University Press, Princeton, New Jersey.

DeVries, P. J. and Walla, T. R. 2001. Species diversity and community structure in neotropical fruit-feeding butterflies. *Biological Journal of the Linnean Society*, **74**, 1–15.

DeVries, P. J., Murray, J. D., and Lande, R. 1997. Species diversity in vertical, horizontal, and temporal dimensions of the fruit-feeding butterfly community in an Ecuadorian rainforest. *Biological Journal of the Linnean Society*, **62**, 343–64.

DeVries, P. J., Walla, T. R., and Greeney, H. F. 1999. Species diversity in spatial and temporal dimensions of fruit-feeding butterflies from two Ecuadorian forests. *Biological Journal of the Linnean Society*, **68**, 333–53.

Dhondt, A. A., Matthysen, E., Adriansen, F., and Lambrechts, M. 1990. Population dynamics and regulation of a high density Blue Tit population. In *Population biology of passerine birds* (eds J. Blondel, A. Gosler, J.-D. Lebreton, and R. McCleery), pp. 39–53. Springer-Verlag, Berlin.

Diffendorfer, J. E., Gaines, M. S., and Holt, R. D. 1999. Patterns and impacts of movements at different scales in small mammals. In *Landscape ecology of small mammals* (eds G. W. Barrett and J. D. Peles), pp. 63–88. Springer-Verlag, New York.

Diserud, O. and Engen, S. 2000. A general and dynamic species abundance model, embracing the lognormal and the gamma models. *American Naturalist*, **155**, 497–511.

Doak, D. 1989. Spotted owls and old growth logging in the Pacific Northwest. *Conservation Biology*, **3**, 389–96.

Doncaster, C. P., Clobert, J., Doligez, B., Gustafsson, L., and Danchin, E. 1997. Balanced dispersal between spatially varying local populations: an alternative to the source-sink model. *American Naturalist*, **150**, 425–55.

Dunnet, G. M., Ollason, J. C., and Anderson, A. 1979. A 28-year study of breeding fulmars *Fulmarus glacialis* in Orkney. *Ibis*, **121**, 293–300.

Dunning, J. B., Jr., Stewart, D. J., Danileson, B. J., Noon, B. R., Root, T. L., Lamberson, R. H., and Stevens, E. E. 1995. Spatially explicit population models: current forms and future uses. *Ecological Applications*, **5**, 3–11.

Durham, J. W. 1970. The fossil record and origin of the *Deuterostomata*. *Proceedings of the North American Paleontology Convention, Chicago 1969*, Section H, pp. 1104–32. University of Chicago Press, Chicago.

Eberhardt, L. L. 1982. Calibrating an index by using removal data. *Journal of Wildlife Management*, **46**, 734–40.

Edwards, P. J., May, R. M., and Webb, N. R. (eds) 1994. *Large-scale ecology and conservation biology*. Blackwell, Oxford.

Efford, M. 2001. Environmental stochasticity can not save declining populations. *Trends in Ecology & Evolution*, **16**, 177.

Efron, B. and Tibshirani, R. J. 1993. *An introduction to the bootstrap*. Chapman & Hall, New York.

Ehrlén, J. and van Groenendael, J. 1998. Direct perturbation analysis for better conservation. *Conservation Biology*, **12**, 470–4.

Ellner, S. and Turchin, P. 1995. Chaos in a noisy world: new methods and evidence from time-series analysis. *American Naturalist*, **145**, 343–75.

Elton, C. S. 1958. *The ecology of invasions by plants and animals*. Methuen, London.

Engen, S. 1975. A note on the geometric series as a species frequency model. *Biometrika*, **62**, 697–9.

Engen, S. 1978. *Stochastic abundance models*. Chapman & Hall, New York.

Engen, S. 2001. A dynamic and spatial model with migration generating the log-Gaussian field of population densities. *Mathematical Biosciences*, **173**, 85–102.

Engen, S. and Lande, R. 1996a. Population dynamic models generating the lognormal species abundance distribution. *Mathematical Biosciences*, **132**, 169–83.

Engen, S. and Lande, R. 1996b. Population dynamic models generating species abundance distributions of the Gamma type. *Journal of Theoretical Biology*, **178**, 325–31.

Engen, S. and Sæther, B.-E. 2000. Predicting the time to quasi-extinction for populations far below their carrying capacity. *Journal of Theoretical Biology*, **205**, 649–58.

Engen, S., Lande, R., and Sæther, B.-E. 1997. Harvesting strategies for fluctuating populations based on uncertain population estimates. *Journal of Theoretical Biology*, **186**, 201–12.

Engen, S., Bakke, Ø., and Islam, A. 1998. Demographic and environmental stochasticity—concepts and definitions. *Biometrics*, **54**, 840–6.

Engen, S., Sæther, B.-E., and Møller, A. P. 2001. Stochastic population dynamics and time to extinction of a declining population of barn swallows. *Journal of Animal Ecology*, **70**, 789–97.

Engen, S., Lande, R., and Sæther, B.-E. 2002a. Migration and spatiotemporal variation in population dynamics in a heterogeneous environment. *Ecology*, **83**, 570–9.

Engen, S., Lande, R., and Sæther, B.-E. 2002b. The spatial scale of population fluctuations and quasi-extinction risk. *American Naturalist*, **160**, 439–51.

Engen, S., Lande, R., Walla, T., and DeVries, P. J. 2002c. Analyzing spatial structure of communities using the two-dimensional Poisson lognormal species abundance model. *American Naturalist*, **160**, 60–73.

Engen, S., Lande, R., and Sæther, B.-E. 2003a. Demographic stochasticity and Allee effects in populations with two sexes. *Ecology* (in press).

Engen, S., Lande, R., and Sæther, B.-E. 2003b. Extinction in relation to demographic and environmental stochasticity in age-structured populations. *American Naturalist* (submitted).

Erb, J. D. and Boyce, M. S. 1999. Distribution of population declines in large mammals. *Conservation Biology*, **13**, 199–201.

Erwin, D. H. 2001. Lessons from the past: Biotic recoveries from mass extinctions. *Proceedings of the National Academy of Sciences, USA*, **98**, 5399–403.

Estes, J. A., Duggins, D. O., and Rathbun, G. B. 1989. The ecology of extinctions in kelp forest communities. *Conservation Biology*, **3**, 252–64.

Euler, L. 1970. A general investigation into the mortality and multiplication of the human species. *Theoretical Population Biology*, **1**, 307–14 (Originally published in 1760).

Ewens, W. J. 1969. *Population genetics*. Methuen, London.

Falconer, D. S. 1965. Maternal effects and selection response. In *Genetics today, Proceedings of the XI International Congress on Genetics*, Vol. 3 (ed. S. J. Geerts), pp. 763–74. Pergamon, Oxford.

Falconer, D. S. and MacKay, T. 1996. *Introduction to quantitative genetics*. 4th edn, Longman, Sussex.

Feldman, M. and Roughgarden, J. 1975. A population's stationary distribution and chance of extinction in a stochastic environment with remarks on the theory of species packing. *Theoretical Population Biology*, **7**, 197–207.

Ferson, S. and Akçakaya, H. R. 1990. *RAMAS/age User Manual*. Applied Biomathematics, Setauket, New York.

Fieberg, J. and Ellner, S. P. 2000. When is it meaningful to estimate an extinction probability? *Ecology*, **81**, 2040–7.

Finlay, B. J., Maberly, S. C., and Cooper, J. I. 1997. Microbial diversity and ecosystem function. *Oikos*, **80**, 209–13.

Fisher, R. A. 1930. *The genetical theory of natural selection*. Oxford at the Clarendon Press, Oxford. (2nd edn, 1958. Dover, New York.)

Fisher, R. A., Corbet, A. S., and Williams, C. B. 1943. The relation between the number of species and the number of individuals in a random sample of an animal population. *Journal of Animal Ecology*, **12**, 42–58.

Foley, P. 1994. Predicting extinction times from environmental stochasticity and carrying capacity. *Conservation Biology*, **8**, 124–37.

Foley, P. 1997. Extinction models for local populations. In *Metapopulation biology. Ecology, genetics, and evolution* (eds I. A. Hanski and M. E. Gilpin), pp. 215–46. Academic Press, San Diego.

Forsman, E. D., DeStefano, S., Raphael, M. G., and Gutiérrez, R. G. 1996. Demography of the northern spotted owl. *Studies in Avian Biology*, **17**, 1–122.

Foster, R. B. 1996. The seasonal rythm of fruitfall on Barro Colorado Island. In *The ecology of a tropical rainforest* (eds E. G. Leigh, Jr., A. S. Rand, and D. M. Windsor), pp. 151–72. Smithsonian Institution Press, Washington, D.C.

Fournier, E. and Loreau, M. 2001. Respective roles of recent hedges and forest patch remnants in the maintenance of ground-beetle (Coleoptera: Carabidae) diversity in an agricultural landscape. *Landscape Ecology*, **16**, 17–32.

Franklin, I. R. 1980. Evolutionary change in small populations. In *Conservation biology, an evolutionary-ecological perspective* (eds M. E. Soulé and B. E. Wilcox), pp. 135–50. Sinauer, Sunderland, Massachusetts.

Gabriel, W. and Bürger, R. 1992. Survival of small populations under demographic stochasticity. *Theoretical Population Biology*, **41**, 44–71.

Gaillard, J.-M., Allainé, D., Pontier, D., Yoccoz, N. G., and Promislow, D. E. L. 1994. Senescence in natural populations of mammals: a reanalysis. *Evolution*, **48**, 509–16.

Gaillard, J.-M., Festa-Bianchet, M., and Yoccoz, N. G. 1998. Population dynamics of large herbivores: variable recruitment with constant adult survival. *Trends in Ecology & Evolution*, **13**, 58–63.

Gaillard, J.-M., Festa-Bianchet, M., Yoccoz, N. G., Loison, A., and Toïgo, C. 2000. Temporal variation in fitness components and population dynamics of large herbivores. *Annual Review of Ecology and Systematics*, **31**, 367–93.

Gaines, M. S. and McClenaghan, L. R., Jr. 1980. Dispersal in small mammals. *Annual Review of Ecology and Systematics*, **11**, 163–96.

Garcia-Charton, J. A. and Perez-Ruzafa, A. 1999. Ecological heterogeneity and the evaluation of the effects of marine reserves. *Fisheries Research*, **42**, 1–20.

Gaston, K. J. and Nicholls, A. O. 1995. Probable times to extinction of some rare breeding bird species in the United Kingdom. *Proceedings of the Royal Society London B, Biological Sciences*, **259**, 119–23.

Gause, G. F. 1934. *The struggle for existence*. Hafner, New York.

Getz, W. M. and Haight, R. G. 1989. *Population harvesting. Demographic models of fish, forest, and animal resources*. Princeton University Press, Princeton.

Gilks, W. R., Richardson, S., and Spiegelhalter, D. J. (eds). 1996. *Markov Chain Monte Carlo in practice*. Chapman & Hall, New York.

Gillespie, J. H. 1991. *The causes of molecular evolution*. Oxford University Press, Oxford.

Gilpin, M. E. 1974. A Liapunov function for competition communities. *Journal of Theoretical Biology*, **44**, 35–48.

Gilpin, M. E. and Ayala, F. J. 1973. Global models of growth and competition. *Proceedings of the National Academy of Sciences, USA*, **70**, 3590–3.

Gilpin, M. E. and Soulé, M. E. 1986. Minimum viable populations: processes of species extinctions. In *Conservation biology. The science of scarcity and diversity* (ed. M. E. Soulé), pp. 13–34. Sinauer, Sunderland, Massachusetts.

Ginzburg, L. R., Slobodkin, L. B., Johnson, K., and Bindman, A. G. 1982. Quasiextinction probabilities as a measure of impact on population growth. *Risk Analysis*, **21**, 171–81.

Goel, N. and Richter-Dyn, N. 1974. *Stochastic models in biology*. Academic Press, New York.

Good, I. J. 1953. The population frequencies of species and the estimation of population parameters. *Biometrika*, **40**, 237–64.

Good, I. J. 1982. Comment. *Journal of the American Statistical Association*, **77**, 561–3.

Goodman, D. 1987. The demography of chance extinction. In *Viable populations for conservation* (ed. M. E. Soulé), pp. 11–34. Cambridge University Press, Cambridge.

Goodman, D. 2002. Predictive Bayesian population viability analysis: a logic for listing criteria, delisting criteria, and recovery plans. In *Population viability analysis* (eds S. Beissinger and D. R. McCullough), pp. 447–69. University of Chicago Press, Chicago.

Goodman, L. A. 1949. On the estimation of the number of classes in a population. *Annals of Mathematical Statistics*, **20**, 572–9.

Goodman, L. A. 1970. The multivariate analysis of qualitative data: interactions among multiple classifications. *Journal of the American Statistical Association*, **65**, 226–56.

Gotelli, N. J. and Graves, G. R. 1996. *Null models in ecology*. Smithsonian Institution Press, Washington, D.C.

Gotelli, N. J. and Kelley, W. G. 1993. A general model of metapopulation dynamics. *Oikos*, **68**, 36–44.

Grassle, J. F. and Smith, W. 1976. A similarity measure sensitive to the contribution of rare species and its use in investigation of variation in marine benthic communities. *Oecologia*, **25**, 13–22.

Grenfell, B. T., Price, O. F., Albon, S. D., and Clutton-Brock, T. H. 1992. Overcompensation and population cycles in an ungulate. *Nature*, **355**, 823–6.

Grenfell, B. T., Wilson, K., Finkenstädt, B. F., Coulson, T. N., Murray, S., Albon, S. D., Pemberton, J. M., Clutton-Brock, T. H., and Crawley, M. J. 1998. Noise and determinism in synchronized sheep dynamics. *Nature*, **394**, 674–7.

Greenman, J. V. and Benton, T. G. 2001. The impact of stochasticity on the behaviour of nonlinear population models: synchrony and the Moran effect. *Oikos*, **93**, 343–51.

Grimm, V. and Wissel, C. 2003. The intrinsic mean time to extinction: a unifying approach to analyzing persistence and viability of populations. (in preparation)

Groom, M. J. 1998. Allee effects limit population viability of an annual plant. *American Naturalist*, **151**, 487–96.

Groom, M. J. and Pascual, M. A. 1998. The analysis of population persistence: an outlook on the practice of viability analysis. In *Conservation biology*. 2nd edn, (eds P. L. Fiedler and P. Kareiva), pp. 4–27. Chapman & Hall, New York.

Groombridge, B. (ed.). 1992. *Global biodiversity. Status of the earth's living resources*. Chapman & Hall, New York.

Grundy, R. M. 1951. The expected frequencies in a sample of an animal population in which the abundances are lognormally distributed. *Biometrika*, **38**, 427–34.

Gulland, J. A. 1971. The effect of exploitation on the numbers of marine mammals. In *Dynamics of populations* (eds P. J. den Boer and G. R. Gradwell), pp. 450–68. Centre for Agricultural Publishing and Documentation, Wageningen.

Gustafsson, L. and Sutherland, W. J. 1988. The costs of reproduction in the collared flycatcher *Ficedula albicollis*. *Nature*, **335**, 813–15.

Hälterlein, B. and Südbeck, P. 1996. Brutbestands-Monitoring von Küstenvögeln an der deutschen Nordseeküste. *Vogelwelt*, **117**, 277–85.

Halliday, T. 1980. *Vanishing birds: their natural history and conservation*. Penguin, Harmondsworth.

Hamilton, W. D. 1966. The moulding of senescence by natural selection. *Journal of Theoretical Biology*, **12**, 12–45.

Hansen, T. F., Stenseth, N. C., Henttonen, H., and Tast, J. 1999. Interspecific and intraspecific competition as causes of direct and delayed density dependence in a fluctuating vole population. *Proceedings of the National Academy of Sciences, USA*, **96**, 986–91.

Hanski, I. 1983. Coexistence of competitors in patchy environment. *Ecology*, **64**, 493–500.

Hanski, I. A. 1999. *Metapopulation ecology*. Oxford University Press, Oxford.

Hanski, I. A. and Gilpin, M. E. (eds). 1997. *Metapopulation biology. Ecology, genetics, and evolution*. Academic Press, San Diego.

Hanski, I. and Gyllenberg, M. 1993. Two general metapopulation models and the core-satellite species hypothesis. *American Naturalist*, **142**, 17–41.

Hanski, I. and Woiwood, I. P. 1993. Spatial synchrony in the dynamics of moth and aphid populations. *Journal of Animal Ecology*, **62**, 656–68.

Hanski, I., Woiwood, I., and Perry, J. 1993. Density dependence, population persistence, and largely futile arguments. *Oecologia*, **95**, 595–8.

Hanson, F. B. and Tuckwell, H. C. 1981. Logistic growth with random density independent disasters. *Theoretical Population Biology*, **19**, 1–18.

Hardin, G. 1968. The tragedy of the commons. *Science*, **162**, 1243–8.

Harms, K. E., Wright, S. J., Caldoron, O., Hernandez, A., and Herre, E. A. 2000. Pervasive density-dependent recruitment enhances seedling diversity in a tropical forest. *Nature*, **404**, 493–5.

Harrison, S. 1993. Species diversity, spatial scale, and global change. In *Biotic interactions and global change* (eds P. M. Kareiva, J. G. Kingsolver, and R. B. Huey), pp. 388–401. Sinauer, Sunderland, Massachusetts.

Harrison, S. and Bruna, E. 1999. Habitat fragmentation and large-scale conservation: what do we know for sure? *Ecography*, **22**, 225–32.

Harrison, S. and Cappuccino, N. 1995. Using density-manipulation experiments to study population regulation. In *Population dynamics: new approaches and synthesis* (eds N. Cappuccino and P. W. Price), pp. 131–47. Academic Press, New York.

Hassell, M. P. and May, R. M. (eds) 1990. *Population regulation and dynamics*. Royal Society, London.

Hassell, M. P., Lawton, J. H., and May, R. M. 1976. Patterns of dynamical behaviour in single-species populations. *Journal of Animal Ecology*, **45**, 471–89.

Hastings, A. 1996. Models of spatial spread: is the theory complete? *Ecology*, **77**, 1675–9.

Hastings, A. and Harrison, S. 1994. Metapopulation dynamics and genetics. *Annual Review of Ecology and Systematics*, **25**, 167–88.

Heck, K. L., van Belle, G., and Simberloff, D. 1975. Explicit calculation of the rarefaction diversity measurement and the determination of sufficient sample size. *Ecology*, **56**, 1459–61.

Hedrick, P. W., Lacy, R. C., Allendorf, F. W., and Soulé, M. R. 1996. Directions in conservation biology: Comments on Caughley. *Conservation Biology*, **10**, 1312–20.

Heino, M., Kaitala, V., Ranta, E., and Lindström, J. 1997. Synchronous dynamics and rates of extinction in spatially structured populations. *Proceedings of the Royal Society London B*, **264**, 481–6.

Heppell, S. S., Walters, J. R., and Crowder, L. B. 1994. Evaluating management alternatives for red-cockaded woodpeckers: a modeling approach. *Journal of Wildlife Management*, **58**, 479–87.

Heyde, C. C. and Cohen, J. E. 1985. Confidence intervals for demographic projections based on products of random matrices. *Theoretical Population Biology*, **27**, 120–53.

Heywood, V. H. (ed.). 1995. *Global biodiversity assessment*. Cambridge University Press, Cambridge.

Higgins, S. I., Pickett, S. T. A., and Bond, W. J. 2000. Predicting extinction risks for plants: environmental stochasticity can save declining populations. *Trends in Ecology & Evolution*, **15**, 516–20.

Hilborn, R. and Walters, C. J. 1992. *Quantitative fisheries stock assessment. Choice, dynamics, and uncertainty*. Chapman & Hall, New York.

Hill, D. 1988. Population dynamics of the avocet (*Recurvirostra avosetta*) breeding in Britain. *Journal of Animal Ecology*, **57**, 669–83.

Hill, M. H. and Caswell, H. 1999. Habitat fragmentation and extinction thresholds on fractal landscapes. *Ecology Letters*, **2**, 121–7.

Hilton-Taylor, C. 2000. *2000 IUCN Red List of Threatened Species*. IUCN, Gland, Switzerland.

Holmes, E. E. 2001. Estimating risks in declining populations with poor data. *Proceedings of the National Academy of Sciences USA*, **98**, 5072–7.

Holmes, E. E. 2002. Beyond theory to application and evaluation: diffusion approximations for population viability analysis. *Ecological Applications* (in review).

Holmes, E. E. and Fagan, W. F. 2002. Validating population viability analysis for corrupted data sets. *Ecology*, **83**, 2379–86.

Hubbell, S. P. 1995. Towards a theory of biodiversity and biogeography on continuous landscapes. In *Preparing for global change: a midwestern perspective* (ed. G. R. Carmichael, G. E. Folk, and J. L. Schnoor), pp. 171–99. SPB Academic, Amsterdam.

Hubbell, S. P. 1997. A unified theory of biogeography and relative species abundance and its application to tropical rainforests and coral reefs. *Coral Reefs*, **16** (Suppl.), S9–21.

Hubbell, S. P. 2001. *The unified neutral theory of biodiversity and biogeography*. Princeton University Press, Princeton.

Huisman, J. and Weissing, F. J. 1999. Biodiversity of plankton by species oscillations and chaos. *Nature*, **402**, 407–10.

Hurlbert, S. H. 1971. The nonconcept of species diversity: a critique and alternative parameters. *Ecology*, **52**, 577–86.

Huston, M. A. 1994. *Biological diversity. The coexistence of species on changing landscapes*. Cambridge University Press, Cambridge.

Hutcheson, K. 1970. A test for comparing diversities based on the Shannon formula. *Journal of Theoretical Biology*, **29**, 151–4.

Hutchings, J. A. 2000. The collapse and recovery of marine fishes. *Nature*, **406**, 882–5.

Hutchings, J. A. and Myers, R. A. 1994. What can be learned from the collapse of a renewable resource—Atlantic cod, *Gadus morhua*, of New Foundland and Labrador? *Canadian Journal of Fisheries and Aquatic Sciences*, **51**, 2126–46.

Hutchinson, G. E. 1957. Concluding remarks. *Cold Spring Harbor Symposium on Quantitative Biology*, **22**, 415–27.

Hutchinson, G. E. 1959. Homage to Santa Rosalia, or why are there so many kinds of animals? *American Naturalist*, **93**, 145–59.

Hutchinson, G. E. 1961. The paradox of the plankton. *American Naturalist*, **95**, 137–45.

Ims, R. A. and Andreassen, H. P. 2000. Spatial synchronization of vole population dynamics by predatory birds. *Nature*, **408**, 194–6.

IUCN 2001. *IUCN Red List categories*. IUCN, Gland, Switzerland.

Järvinen, A. 1990. Changes in the abundance of birds in relation to small rodent density and predation rate in Finnish Lapland. *Bird Study*, **37**, 36–9.

Jablonski, D. 1986. Background and mass extinctions—the alternation of macroevolutionary regimes. *Science*, **231**, 129–33.

Jablonski, D. 1989. The biology of mass extinction: a palaeontological view. *Philosophical Transactions of the Royal Society B*, **325**, 357–68.

Jablonski, D. 1991. Extinctions: a paleontological perspective. *Science*, **253**, 754–7.

Jablonski, D. 1994. Extinctions in the fossil record. *Philosophical Transactions of the Royal Society B*, **344**, 11–17.

Jackson, J. B. C., Kirby, M. X., Berger, W. H., Bjorndal, K. A., Botsford, L. W., Bourque, B. J., Bradbury, R. H., Cooke, R., Erlandson, J., Estes, J. A., Hughes, T. P., Kidwell, S., Lange, C. B., Lenihan, H. S., Pandolfi, J. M., Peterson, C. H., Steneck, R. S., Tegner, M. J., and Warner, R. R. 2001. Historical overfishing and the recent collapse of coastal ecosystems. *Science*, **293**, 629–38.

James, A. N., Gaston, K. J., and Balmford, A. 1999. Balancing the Earth's accounts. *Nature*, **401**, 323–4.

James, F. C. 1995. Prehistoric extinctions and ecological changes on oceanic islands. In *Islands: Biological diversity and ecosystem function* (eds P. M. Vitousek, L. L. Loope, and H. Adsersen), pp. 87–102. Springer-Verlag, New York.

Janzen, D. H. 1970. Herbivores and the number of tree species in tropical forests. *American Naturalist*, **104**, 501–28.

Jensen, A. L. 1999. Using simulation to verify life history relations indicated by time series analysis. *Environmetrics*, **10**, 237–45.

Johnson, N. L. and Kotz, S. 1970. *Distributions in statistics: continuous univariate distributions*, Vol. 1. Wiley, New York.

Johst, K. and Wissel, C. 1997. Extinction risk in a temporally correlated fluctuating environment. *Theoretical Population Biology*, **52**, 91–100.

Jouventin, P., and Guillotin, M. 1979. Socioécologie du skua antarctique à Ponte géologie. *Terre et Vie*, **38**, 109–27.

Kaitala, V., Ylikarjula, J., Ranta, E., and Lundberg, P. 1997. Population dynamics and the colour of environmental noise. *Proceedings of the Royal Society of London B*, **264**, 943–8.

Kareiva, P. and Wennergren, U. 1995. Connecting landscape patterns to ecosystem and population processes. *Nature*, **373**, 299–302.

Karlin, S. and Taylor, H. M. 1975. *A first course in stochastic processes*. Academic Press, New York.

Karlin, S. and Taylor, H. M. 1981. *A second course in stochastic processes*. Academic Press, New York.

Keeling, M. J., Woolhouse, M. E. J., Shaw, D. J., Matthews, L., Chase-Topping, M., Haydon, D. T., Cornell, S. J., Kappey, J., Wilesmith, J., and Grenfell, B. T. 2001. Dynamics of the 2001 UK foot and mouth epidemic: stochastic dispersal in a heterogeneous landscape. *Science*, **294**, 813–17.

Kelly, C. K. and Bowler, M. G. 2002. Coexistence and relative abundance in forest trees. *Nature*, **417**, 437–40.

Kempton, R. A. 1979. The structure of species abundance and measurement of diversity. *Biometrics*, **35**, 307–21.

Kempton, R. A and Taylor, L. R. 1974. Log-series and log-normal parameters as diversity discriminants for the Lepidoptera. *Journal of Animal Ecology*, **43**, 381–99.

Kendall, B. E., Bjørnstad, O. N., Bascompte, J., Keitt, T. H., and Fagan, W. F. 2000. Dispersal, environmental correlation, and spatial synchrony in population dynamics. *American Naturalist*, **155**, 628–36.

Kendall, M. G., Stuart, A., and Ord, J. K. 1983. *The advanced theory of statistics*, Vol. 3, *Design and analysis, and time series*. 5th edn, Charles Griffin and Company, London.

Keyfitz, N. 1968. *Introduction to the mathematics of population*. Addison-Wesley, Reading, Massachusetts.

Kierstead, H. and Slobodkin, L. R. 1953. The sizes of water masses containing plankton blooms. *Journal of Marine Research*, **12**, 141–7.

Kiester, A. R. 1979. Conspecifics as cues: a mechanism for habitat selection in the Panamanian grass anole (*Anolis auratus*). *Behavioral Ecology and Sociobiology*, **5**, 323–30.

Kimura, M. 1983. *The neutral theory of molecular evolution*. Cambridge University Press, Cambridge.

Kinzig, A. P., Pacala, S. P., and Tilman, D. 2002. *The functional consequences of biodiversity*. Princeton University Press, Princeton.

Kirkpatrick, M. and Lande, R. 1989. The evolution of maternal characters. *Evolution*, **43**, 485–503.

Koenig, W. D. 1998. Spatial autocorrelation in California land birds. *Conservation Biology*, **12**, 612–20.

Komdeur, J. 1994. Conserving the Seychelles warbler *Acrocephalus sechellensis* by translocation from Cousin Island to the islands of Aride and Cousine. *Biological Conservation*, **67**, 143–52.

Kot, M., Lewis, M. A., and van den Driessche, P. 1996. Dispersal data and the spread of invading organisms. *Ecology*, **77**, 2027–42.

Korn, G. A. and Korn, T. M. 1968. *Mathematical handbook for scientists and engineers*. 2nd edn, McGraw-Hill, New York.

Lacy, R. G. 1993. VORTEX—a computer-simulation model for population viability analysis. *Wildlife Research*, **20**, 45–65.

Lacy, R. G. and Miller, P. S. 2002. Incorporating human populations and activities into population viability analysis. In *Population viability analysis* (eds S. Beissinger and D. R. McCullough), pp. 490–510. University of Chicago Press, Chicago.

Lamberson, R. H., McKelvey, R., Noon, B. R., and Voss, C. 1992. A dynamic analysis of Northern Spotted Owl viability in a fragmented landscape. *Conservation Biology*, **6**, 505–12.

Lamberson, R. H., Noon, B. R., Voss, C., and McKelvey, K. S. 1994. Reserve design for territorial species: the effects of patch size and spacing on the viability of the Northern Spotted Owl. *Conservation Biology*, **8**, 185–95.

Lande, R. 1982*a*. A quantitative genetic theory of life history evolution. *Ecology*, **63**, 607–15.

Lande, R. 1982*b*. Elements of a quantitative genetic model of life history evolution. In *Evolution and genetics of life histories* (eds H. Dingle and J. P. Hegmann), pp. 21–9. Springer-Verlag, New York.

Lande, R. 1987. Extinction thresholds in demographic models of territorial species. *American Naturalist*, **130**, 624–35.

Lande, R. 1988. Demographic models of the northern spotted owl (*Strix occidentalis caurina*). *Oecologia*, **75**, 601–7.

Lande, R. 1993. Risks of population extinction from demographic and environmental stochasticity and random catastrophes. *American Naturalist*, **142**, 911–27.

Lande, R. 1995. Mutation and conservation. *Conservation Biology*, **9**, 782–91.

Lande, R. 1996. Statistics and partitioning of species diversity, and similarity among multiple communities. *Oikos*, **76**, 5–13.

Lande, R. 1998. Demographic stochasticity and Allee effect on a scale with isotrophic noise. *Oikos*, **83**, 353–8.

Lande, R. 2002. Incorporating stochasticity in population viability analysis. In *Population viability analysis* (eds S. Beissinger and D. R. McCullough), pp. 18–40. University of Chicago Press, Chicago.

Lande, R. and Orzack, S. H. 1988. Extinction dynamics of age-structured populations in a fluctuating environment. *Proceedings of the National Academy of Sciences, USA*, **85**, 7418–21.

Lande, R. and Shannon, S. 1996. The role of genetic variability in adaptation and population persistence in a changing environment. *Evolution*, **50**, 434–7.

Lande, R., Engen, S., and Sæther, B.-E. 1995. Optimal harvesting of fluctuating populations with a risk of extinction. *American Naturalist*, **145**, 728–45.

Lande, R., Sæther, B.-E., and Engen, S. 1997. Threshold harvesting for sustainability of fluctuating resources. *Ecology*, **78**, 1341–50.

Lande, R., Engen, S., and Sæther, B.-E. 1998. Extinction times in finite metapopulation models with stochastic local dynamics. *Oikos*, **83**, 383–9.

Lande, R., Engen, S., and Sæther, B.-E. 1999. Spatial scale of population synchrony: environmental correlation versus dispersal and density regulation. *American Naturalist*, **154**, 271–81.

Lande, R., DeVries, P. J., and Walla, T. R. 2000. When species accumulation curves intersect: implications for ranking diversity using small samples. *Oikos*, **89**, 601–5.

Lande, R., Sæther, B.-E., and Engen, S. 2001. Sustainable exploitation of fluctuating populations. In *Conservation of exploited species* (eds J. D. Reynolds, G. M. Mace, K. H. Redford and J. G. Robinson), pp. 67–86. Cambridge University Press, Cambridge.

Lande, R., Sæther, B.-E., Engen, S., Filli, F., Matthysen, E., and Weimerskirch, H. 2002. Estimating density dependence from population time series using demographic theory and life-history data. *American Naturalist*, **159**, 321–32.

Larkin, P. A. 1977. An epitaph for the concept of maximum sustained yield. *Transactions of the American Fisheries Society*, **106**, 1–11.

Larsen, D. G., Gauthier, D. A., and Markel, R. L. 1989. Causes and rate of moose mortality in the southwest Yukon. *Journal of Wildlife Management*, **53**, 548–57.

Le Boef, B. J. 1981. Mammals. In *The natural history of Año Nuevo* (eds B. J. Le Boef and S. Kaza), pp. 287–325. Boxwood, Pacific Grove, California.

Lee, D. C. 2000. Assessing land-use impacts on Bull Trout using Bayesian belief networks. In *Quantitative methods for conservation biology* (eds S. Ferson and M. Burgman), pp. 127–47. Springer-Verlag, New York.

Lefkovitch, L. P. 1965. The study of population growth in organisms grouped by stages. *Biometrics*, **21**, 1–18.

Leigh, E. G., Jr. 1981. The average lifetime of a population in a varying environment. *Journal of Theoretical Biology*, **90**, 213–39.

Leigh, E. G., Jr. 1999. *Tropical forest ecology. A view from Barro Colorado Island*. Oxford University Press, Oxford.

Leslie, P. H. 1945. On the use of matrices in certain population mathematics. *Biometrika*, **33**, 183–212.

Leslie, P. H. 1948. Some further notes on the use of matrices in population mathematics. *Biometrika*, **35**, 213–45.

Leslie, P. H. 1966. The intrinsic rate of increase and the overlap of successive generations in a population of guillemots (*Uria aalge* Pont.) *Journal of Animal Ecology*, **35**, 291–301.

Levin, S. A. 1970. Community equilibria and stability, and an extension of the competitive exclusion principle. *American Naturalist*, **104**, 413–23.

Levins, R. 1968. *Evolution in changing environments*. Princeton University Press, Princeton.

Levins, R. 1969. Some demographic and genetic consequences of environmental heterogeneity for biological control. *Bulletin of the Entomological Society of America*, **15**, 237–40.

Levins, R. 1970. Extinction. In *Some mathematical questions in biology* (ed. M. Gerstenhaber), pp. 75–107. The American Mathematical Society, Providence, Rhode Island.

Levins, R. 1979. Coexistence in a variable environment. *American Naturalist*, **114**, 765–83.

Levitan, D. R. and Petersen, C. 1995. Sperm limitation in the sea. *Trends in Ecology & Evolution*, **10**, 228–31.

Lewis, M. A. and Kareiva, P. 1993. Allee dynamics and the spread of invading organisms. *Theoretical Population Biology*, **43**, 141–58.

Lewontin, R. C. 1972. The apportionment of human diversity. *Evolutionary Biology*, **6**, 381–98.

Lewontin, R. C. and Cohen, D. 1969. On population growth in a randomly varying environment. *Proceedings of the National Academy of Sciences, USA*, **62**, 1056–60.

Lichatowich, J. 1999. *Salmon without rivers*. Island Press, Washington, D.C.

Linden, E. 1994. Tigers on the brink. *Time*, **143**, 44–51.

Lindenmayer, D. B. and Lacy, R. C. 2002. Small mammals, habitat patches and PVA models: a field test of model predictive ability. *Biological Conservation*, **103**, 247–65.

Lindenmayer, D. B., Burgman, M. A., Akçakaya, H. R., Lacy, R. C., and Possingham, H. P. 1995. A review of the generic computer programs ALEX, RAMAS/space and VOR-TEX for modeling the viability of wildlife metapopulations. *Ecological Modelling*, **82**, 161–74.

Lindenmayer, D. B., Lacy, R. L., and Pope, M. L. 2000. Testing a simulation model for population viability analysis. *Ecological Applications*, **10**, 580–97.

Lindenmayer, D. B., Ball, I., Possingham, H. P., McCarthy, M. A., and Pope, M. L. 2001. A landscape-scale test of the predictive ability of a spatially explicit model for population viability analysis. *Journal of Applied Ecology*, **38**, 36–48.

Lindström, J., Ranta, E., and Lindén, H. 1996. Large-scale synchrony in the dynamics of the capercaillie, black grouse and hazel grouse populations in Finland. *Oikos*, **76**, 221–7.

Loison, A., Gaillard, J.-M., and Houssin, H. 1994. New insight on survivorship of female chamois (*Rupicapra rupicapra*) from observations of marked animals. *Canadian Journal of Zoology*, **72**, 591–7.

Loison, A., Festa-Bianchet, M., Gaillard, J.-M., Jorgenson, J. T., and Julien, J.-M. 1999. Age-specific survival in five populations of ungulates: Evidence of senescence. *Ecology*, **80**, 2539–54.

Lotka, A. J. 1924. *Elements of physical biology*. Williams & Watkins, Baltimore. (Reprinted and revised as *Elements of mathematical biology*, 1956. Dover, New York).

Lowe, S. A., and Thompson, G. G. 1993. Accounting for uncertainties in the development of exploitation strategies for Atca mackerel resource of the Aleutian Islands. In *Management strategies for exploited fish populations* (eds G. Kruse, D. M. Eggers, R. J. Marasco, C. Pautzke, and J. T. Quinn II), pp. 203–31. Alaska Sea Grant College Program, University of Alaska, Fairbanks.

Ludwig, D. 1975. Persistence of dynamical systems under random perturbations. *SIAM Review*, **17**, 605–40.

Ludwig, D. 1976. A singular perturbation problem in the theory of population extinction. *Society for Industrial and Applied Mathematics—American Mathematical Society Proceedings*, **10**, 87–104.

Ludwig, D. 1996*a*. The distribution of population survival times. *American Naturalist*, **147**, 506–26.

Ludwig, D. 1996*b*. Uncertainty and the assessment of extinction probabilities. *Ecological Applications*, **6**, 1067–76.

Ludwig, D. 1999. Is it meaningful to estimate a probability of extinction? *Ecology*, **80**, 298–310.

Ludwig, D., Hilborn, R., and Walters, C. 1993. Uncertainty, resource exploitation, and conservation: lessons from history. *Science*, **260**, 17–18.

Lynch, M. and Walsh, B. 1998. *Genetics and analysis of quantitative traits*. Sinauer Associates, Sunderland, Massachusetts.

MacArthur, R. H. 1960. On the relative abundance of species. *American Naturalist*, **94**, 25–36.

MacArthur, R. H. 1965. Patterns of species diversity. *Biological Reviews*, **40**, 510–33.

MacArthur, R. H. 1972. *Geographical ecology*. Harper & Row, New York.

MacArthur, R. H. and Wilson, E. O. 1967. *The theory of island biogeography*. Princeton Univeristy Press, Princeton.

MacDonald, N. 1978. *Time lags in biological models*. Springer-Verlag, Berlin.

Mace, G. M. and Lande, R. 1991. Assessing extinction threats: toward a reevaluation of IUCN threatened species categories. *Conservation Biology*, **5**, 148–57.

MacPhee, R. D. E. (ed.). 1999. *Extinctions in near time*. Kluwer Academic/Plenum Publishing, New York.

Magurran, A. E. 1988. *Ecological diversity and its measurement*. Princeton University Press, Princeton.

Mangel, M. 2000. Irreducible uncertainties, sustainable fisheries and marine reserves. *Evolutionary Ecology Research*, **2**, 547–57.

Mangel, M. and Tier, C. 1993. Dynamics of metapopulations with demographic stochasticity and environmental catastrophes. *Theoretical Population Biology*, **44**, 1–31.

Manly, B. F. J. 1991. *Randomization and Monte Carlo methods in biology*. Chapman & Hall, New York.

Marcus, M. and Minc, H. 1964. *A survey of matrix theory and matrix inequalities*. Dover, New York.

Martin, P. S. 1984. In *Quaternary extinctions: a prehistoric revolution* (eds P. S. Martin and R. G. Klein). University of Arizona Press, Tucson.

Martin, P. S. 1986. Refuting Late Pleistocene extinction models. In *Dynamics of extinction* (ed. D. K. Elliott), pp. 107–30. Wiley, New York.

Martin, P. S. and Klein, R. G. (eds). 1984. *Quaternary extinctions: a prehistoric revolution*. University of Arizona Press, Tucson.

Martin, P. S. and Steadman, D. W. 1999. Prehistoric extinctions on islands and continents. In *Extinctions in near time* (ed. R. D. E. MacPhee), pp. 17–55. Kluwer Academic/Plenum Publishing, New York.

May, R. M. 1973. Stability in randomly fluctuating versus deterministic environments. *American Naturalist*, **107**, 621–50.

May, R. M. 1974. *Stability and complexity in model ecosystems*. 2nd edn, Princeton Univeristy Press, Princeton.

May, R. M. 1975. Patterns of species abundance and diversity. In *Ecology and evolution of communities* (eds M. L. Cody and J. L. Diamond), pp. 81–120. Harvard University Press, Cambridge, Massachusetts.

May, R. M. 1976. Simple mathematical models with very complicated dynamics. *Nature*, **261**, 459–67.

May, R. M. 1981. Models for single populations. In *Theoretical ecology*. 2nd edn (ed. R. M. May), pp. 5–29. Blackwell Scientific Publications, Oxford.

May, R. M. and Oster, G. F. 1976. Bifurcations and dynamic complexity in simple ecological models. *American Naturalist*, **110**, 573–99.

May, R. M., Lawton, J. H., and Stork, N. E. 1995. Assessing extinction rates. In *Extinction rates* (eds J. H. Lawton and R. M. May), pp. 1–24. Oxford University Press, Oxford.

McCarty, J. P. 2001. Ecological consequences of recent climate changes. *Conservation Biology*, **15**, 320–31.

McCarthy, M. A., Burgman, M. A., and Ferson, S. 1995. Sensitivity analysis for models of population viability. *Biological Conservation*, **73**, 93–100.

McCullagh, P. and Nelder, J. A. 1989. *Generalized linear models*. 2nd edn, Chapman & Hall, New York.

McKelvey, K., Noon, B. R., and Lamberson, R. H. 1993. Conservation planning for species occupying fragmented landscapes: the case of the Northern Spotted Owl. In *Biotic interactions and global change* (eds P. M. Kareiva, J. G. Kingsolver, and R. B. Huey), pp. 424–50. Sinauer, Sunderland, Massachusetts.

McKinney, M. L. 1998. Branching models predict loss of many bird and mammal orders within centuries. *Animal Conservation*, **1**, 159–64.

Mertz, D. B. 1971. The mathematical demography of the California condor population. *American Naturalist*, **105**, 437–53.

Middleton, D. A. J. and Nisbet, R. M. 1997. Population persistence time: estimates, models, and mechanisms. *Ecological Applications*, **7**, 107–17.

Miller, G. T., Jr. 1990. *Living in the environment. An introduction to environmental science*. 6th edn, Wadsworth Publishing Co., Belmont, California.

Mills, L. S., Hayes, S. G., Baldwin, C., Wisdom, M. J., Citta, J., Mattson, D. J., and Murphy, K. 1996. Factors leading to different viability predictions for a grizzly bear data set. *Conservation Biology*, **10**, 863–73.

Møller, A. P. 1989. Population dynamics of a declining swallow *Hirundo rustica* population. *Journal of Animal Ecology*, **58**, 1051–63.

Møller, A. P. 2001. The effect of dairy farming on barn swallow *Hirundo rustica* abundance, distribution and reproduction. *Journal of Applied Ecology*, **38**, 378–89.

Moran, P. A. P. 1953. The statistical analysis of the Canadian lynx cycle. II. Synchronization and meteorology. *Australian Journal of Zoology*, **1**, 291–8.

Mosqueira, I., Côte, I. M., Jennings, S., and Reynolds, J. D. 2000. Conservation benefits of marine reserves for fish populations. *Animal Conservation*, **3**, 321–32.

Motomura, I. 1932. A statistical treatment of associations [in Japanese]. *Japanese Journal of Zoology*, **44**, 379–83.

Murdoch, W. W. 1994. Population regulation in theory and practice. *Ecology*, **75**, 271–87.

Myers, J. H. 1998. Synchrony in outbreaks of forest lepidoptera: a possible example of the Moran effect. *Ecology*, **79**, 1111–17.

Myers, R. A., Barrowman, N. J., Hutchings, J. A., and Rosenberg, A. A. 1995. Population dynamics of exploited species at low population levels. *Science*, **269**, 1106–8.

Myers, R. A., Mertz, G. A., and Bridson, J., 1997a. Spatial scales of interannual recruitment variations of marine, anadromous, and freshwater fish. *Canadian Journal of Fisheries and Aquatic Sciences*, **54**, 1400–7.

Myers, R. A., Hutchings, J. A., and Barrowman, N. J. 1997b. Why do fish stocks collapse? The example of cod in Atlantic Canada. *Ecological Applications*, **7**, 91–106.

Myers, R. A., Bowen, K. G., and Barrowman, N. J., 1999. Maximum reproductive rate of fish at low densities. *Canadian Journal of Fisheries and Aquatic Sciences*, **56**, 2404–19.

Namias, J. 1978. Persistence of U. S. seasonal temperatures up to one year. *Monthly Weather Review*, **106**, 1557–67.

Nåsell, I. 2001. Extinction and quasi-stationarity in the Verhulst logistic model. *Journal of Theoretical Biology*, **211**, 11–27.

Naeh, T., Klosek, M. M., Matkowsky, B. J., and Schuss, Z. 1990. A direct approach to the exit problem. *SIAM Journal of Applied Mathematics*, **50**, 595–627.

Nee, S. 1994. How populations persist. *Nature*, **367**, 123–4.

Nei, M. 1973. Analysis of gene diversity in subdivided populations. *Proceedings of the National Academy of Sciences USA*, **70**, 3321–3.

Newton, I. 1989. *Lifetime reproduction in birds*. Academic Press, London.

Nichols, J. D., Hines, J. E., and Blums, P. 1997. Tests for senescent decline in annual survival probabilities of common poachards, *Aythya ferina*. *Ecology*, **78**, 1009–18.

Nicholls, N. 1980. Long range weather forecasting: value, status, and prospects. *Reviews of Geophysics and Space Physics*, **18**, 771–88.

Nisbet, R. M. 1997. Delay-differential equations for structured populations. In *Structured-population models in marine, terrestrial, and freshwater systems* (eds S. Tuljapurkar and H. Caswell), pp. 89–118. Chapman & Hall, New York.

Nisbet, R. M. and Gurney, W. S. C. 1982. *Modelling fluctuating populations*. Wiley, New York.

Nobile, A. G., Ricciardi, L. M., and Sacerdote, L. 1985. Exponential trends of first-passage time densities for a class of diffusion processes with steady-state distributions. *Journal of Applied Probability*, **22**, 611–18.

Noon, B., McKelvey, K., and Murphy, D. 1997. Developing an analytical context for multi-species conservation planning. In *The ecological basis of conservation. Heterogeneity, ecosystems, and biodiversity* (eds S. T. A. Pickett, R. S. Ostfeld, M. Shachak, and G. E. Likens), pp. 43–59. Chapman & Hall, New York.

North, P. M. and Morgan, B. J. T. 1979. Modelling heron survival using weather data. *Biometrics*, **35**, 667–81.

Okubo, A. 1980. *Diffusion and ecological problems: mathematical models*. Springer-Verlag, New York.

O'Riordan, T. and Cameron, J. 1995. *Interpreting the precautionary principle*. Earthscan Publications, London.

O'Riordan, T., Cameron, J., and Jordan A. (eds). 2001. *Reinterpreting the precautionary principle*. Cameron May, London.

Orzack, S. H. 1997. Life history evolution and extinction. In *Structured-population models in marine, terrestrial, and freshwater systems* (eds S. Tuljapurkar and H. Caswell), pp. 273–302. Chapman & Hall, New York.

Overholtz, W. J., Edwards, S. F., and Brodziak, J. K. T. 1993. Strategies for rebuilding and harvesting New England groundfish resources. In *Management strategies for exploited fish populations* (eds G. Kruse, D. M. Eggers, R. J. Marasco, C. Pautzke, and J. T. Quinn II), pp. 507–27. Alaska Sea Grant College Program, University of Alaska, Fairbanks.

Owen, D. F. 1960. The nesting success of the heron *Ardea cinerea* in relation to the availability of food. *Proceedings of the Zoological Society, London*, **133**, 597–617.

Owen-Smith, R. N. 1988. *Megaherbivores: the influence of very large body size on ecology.* Cambridge University Press, Cambridge.

Palmer, M. W. 1990. The estimation of species richness by extrapolation. *Ecology*, **71**, 1195–8.

Palmer, M. W. 1991. Estimating species richness: the second-order jackknife reconsidered. *Ecology*, **72**, 1512–13.

Palmqvist, E. and Lundberg, P. 1998. Population extinction in correlated environments. *Oikos*, **83**, 359–67.

Paradis, E., Baillie, S. R., Sutherland, W. J., and Gregory, R. D. 2000. Spatial synchrony in populations of birds: effects of habitat, population trend, and spatial scale. *Ecology*, **81**, 2112–25.

Park, T. 1957. Experimental studies of interspecies competition. III. Relation of initial species proportion to competitive outcome in populations of Tribiolum. *Physiological Zoology*, **30**, 22–40.

Parker, W. C. and Arnold, A. J. 1997. Species survivorship in the Cenozoic planktonic foraminifera: a test of exponential and Weibull models. *Palaios*, **12**, 3–11.

Parr, R. 1992. The decline to extinction of a population of Golden Plover in north-east Scotland. *Ornis Scandinavica*, **23**, 152–8.

Patil, G. P. and Taillie, C. 1982. Diversity as a concept and its measurement. *Journal of the American Statistical Association*, **77**, 548–61.

Patrick, R. 1963. The structure of diatom communities under varying ecological conditions. *Annals of the New York Academy of Sciences*, **108**, 359–65.

Pease, C. M., Lande, R., and Bull, J. J. 1989. A model of population growth, dispersal and evolution in a changing environment. *Ecology*, **70**, 1657–64.

Pech, R. P. and Hood, G. M. 1998. Foxes, rabbits, alternative prey and rabbit calciviris disease: consequences of a new biological control agent for an outbreaking species in Australia. *Journal of Applied Ecology*, **35**, 434–53.

Pech, R. P., Hood, G. M., Singleton, G. R., Salmon, E., Forrester, R. I., and Brown, P. R. 1999. Models for predicting plagues of house mice (*Mus domesticus*) in Australia. In *Ecologically-based management of rodent pests* (eds G. Singleton, H. Leirs, Z. Zhang, and L. Hinds), pp. 81–112. Australian Centre for International Agricultural Research, Canberra.

Peet, R. K. 1974. The measurement of species diversity. *Annual Review of Ecology and Systematics*, **5**, 285–307.

Peters, R. L. and Lovejoy, T. E. 1992. *Global warming and biological diversity.* Yale University Press, New Haven.

Pfister, C. M. 1998. Patterns of variance in stage-structured populations: Evolutionary predictions and ecological implications. *Proceedings of the National Academy of Sciences, USA*, **95**, 213–18.

Pielou, E. C. 1966. Shannon's formula as a measure of specific diversity: its use and misuse. *American Naturalist*, **100**, 463–5.

Pielou, E. C. 1969. *An introduction to mathematical ecology.* Wiley, New York.

Pielou, E. C. 1975. *Ecological diversity.* Wiley, New York.

Pimm, S. L. 1991. *The balance of nature?* University of Chicago Press, Chicago.

Pimm, S. L. and Redfearn, A. 1988. The variability of animal populations. *Nature*, **334**, 613–14.

Pimm, S. L., Russell, G. J., Gittleman, J. L., and Brooks, T. M. 1995. The future of biodiversity. *Science*, **269**, 347–50.

Pollard, E., Lakhani, K. H., and Rothery, P. 1987. The detection of density-dependence from a series of annual censuses. *Ecology*, **68**, 2046–53.

Pope, J. G., Macdonald, D. S., Daan, N., Reynolds, J. D., and Jennings, S. 2000. Guaging the impact of fishing mortality on non-target species. *ICES Journal of Marine Science*, **57**, 689–96.

Possingham, H. P. and Davies, I. 1995. ALEX: a model for the viability analysis of spatially structured populations. *Biological Conservation*, **73**, 143–50.

Post, E. and Stenseth, N. C. 1999. Climate change, plant phenology, and northern ungulates. *Ecology*, **80**, 1322–39.

Poulin, R. 1998. *Evolutionary ecology of parasites: from individuals to communities*. Chapman & Hall, New York.

Preston, F. W. 1948. The commonness, and rarity, of species. *Ecology*, **29**, 254–83.

Preston, F. W. 1962. The canonical distribution of commonness and rarity. *Ecology*, **43**, 185–215, 410–32.

Preston, F. W. 1980. Noncanonical distributions of commonness and rarity. *Ecology*, **61**, 88–97.

Primack, R. B. 1993. *Essentials of conservation biology*. Sinauer, Sunderland, Massachusetts.

Pulliam, H. R. 1988. Sources, sinks, and population regulation. *American Naturalist*, **107**, 652–61.

Pulliam, H. R. and Danielson, B. J. 1992. Sources, sinks, and habitat selection: a landscape perspective on population dynamics. *American Naturalist*, **137**, S50–66.

Purvis, A, Jones, K. E., and Mace, G. M. 2000. Extinction. *BioEssays*, **22**, 1123–33.

Quinn, T. J. II and Deriso, R. B. 1999. *Quantitative fish dynamics*. Oxford University Press, New York.

Ranta, E., Kaitala, V., and Lundberg, P. 1997. The spatial dimension in population fluctuations. *Science*, **278**, 1621–3.

Ranta, E., Kaitala, V., and Lindström, J. 1998. Spatial dynamics of populations. In *Modeling Spatiotemporal dynamics in ecology* (eds J. Bascompte and R. V. Solé), pp. 47–62. Springer-Verlag, Berlin.

Rao, C. R. 1984. Convexity properties of entropy functions and analysis of diversity. In *Inequalities in statistics and probability* (ed. Y. L. Tong). IMS Lecture Notes—Monograph Series Vol. 5, pp. 68–77.

Ratner, S., Lande, R., and Roper, B. B. 1997. Population viability analysis of spring chinook salmon in the South Umpqua River, Oregon. *Conservation Biology*, **11**, 879–89.

Raup, D. M. and Stanley, S. M. 1978. *Principles of paleontology*. 2nd edn, Freeman, San Fransisco.

Redford, K. H. 1992. The empty forest. *Bioscience*, **42**, 412–22.

Redford, K. H., Coppolillo, P., Sanderson, E. W., da Fonseca, G. A. B., Dinerstein, E., Groves, C., Mace, G., Maginnis, S., Mittermeier, R. A., Noss, R., Olson, D., Robinson, J. G., Vedder, A., and Wright, M. 2002. Mapping the conservation landscape. *Conservation Biology* (in press).

Reed, W. J. 1978. The steady state of a stochastic harvesting model. *Mathematical Biosciences*, **41**, 273–307.

Reed, W. J. 1979. Optimal escapement levels in stochastic and deterministic harvesting models. *Journal of Environmental Economics and Management*, **6**, 350–63.

Renshaw, E. 1991. *Modelling biological populations in space and time*. Cambridge University Press, Cambridge.

Richter-Dyn, N. and Goel, N. S. 1972. On the extinction of a colonizing species. *Theoretical Population Biology*, **3**, 406–33.

Ringsby, T. H., Sæther, B.-E., Altwegg, R., and Solberg, E. J. 1999. Temporal and spatial variation in survival rates of a house sparrow, *Passer domesticus*, metapopulation. *Oikos*, **85**, 419–25.

Ripa, J. and Heino, M. 1999. Linear analysis solves two puzzles in population dynamics: the route to extinction and extinction in coloured environments. *Ecology Letters*, **2**, 219–22.

Ripa, J. and Lundberg, P. 1996. Noise colour and the risk of population extinctions. *Proceedings of the Royal Society London B*, **263**, 1751–3.

Ripa, J., Lundberg, P., and Kaitala, V. 1998. A general theory of environmental noise in ecological food webs. *American Naturalist*, **151**, 256–63.

Robbins, C. S., Dawson, D. K., and Dowell, B. A. 1989. Habitat area requirements of breeding forest birds of the middle Atlantic states. *Wildlife Monographs*, **103**, 1–34.

Roberts, L. 1991. Ranking the rain-forests. *Science*, **251**, 1559–60.

Romer, A. S. 1966. *Vertebrate paleontology*. 3rd edn, University of Chicago Press, Chicago.

Rosenberg, A. A., Fogarty, M. J., Sissenwine, M. P., Beddington, J. R., and Shepherd, J. G. 1993. Achieving sustainable use of renewable resources. *Science*, **262**, 828–9.

Rosenzweig, M. L. 1995. *Species diversity in space and time*. Cambridge University Press, Cambridge.

Routledge, R. D. 1977. On Whittaker's components of diversity. *Ecology*, **58**, 1120–7.

Roy, K., Valentine, J. W., Jablonski, D., and Kidwell, S. M. 1996. Scales of climatic variability and time averaging in Pleistocene biotas: Implications for ecology and evolution. *Trends in Ecology & Evolution*, **11**, 458–63.

Royama, T. 1992. *Analytical population dynamics*. Chapman & Hall, New York.

RSRP. 2001. Salmon Recovery Science Review Panel: Report for meeting held August 27–29, 2001, National Marine Fisheries Service, Northwest Fisheries Science Center, Seattle, Washington. http://www.nwfsc.noaa.gov/cbd/trt/rsrp.htm

Sæther, B.-E. 1985. Annual variation in carcass weight of Norwegian moose in relation to climate along a latitudinal gradient. *Journal of Wildlife Management*, **49**, 977–83.

Sæther, B.-E. 1990. Age-specific variation in reproductive performance of birds. *Current Ornithology*, **7**, 251–83.

Sæther, B.-E. 1997. Environmental stochasticity and population dynamics of large herbivores: a search for mechanisms. *Trends in Ecology & Evolution*, **12**, 143–9.

Sæther, B.-E. and Bakke, Ø. 2000. Avian life history variation and contribution of demographic traits to the population growth rate. *Ecology*, **81**, 642–53.

Sæther, B.-E. and Engen, S. 2002. Including uncertainties in population viability analysis using population prediction intervals. In *Population viability analysis* (eds S. Beissinger and D.R. McCullough), pp. 191–212. University of Chicago Press, Chicago.

Sæther, B.-E. and Heim, M. 1993. Ecological correlates of individual variation in age at maturity in female moose (*Alces alces*): the effects of environmental variability. *Journal of Animal Ecology*, **62**, 482–9.

Sæther, B.-E., Andersen, R., Hjeljord, O., and Heim, M. 1996a. Ecological correlates of regional variation in life history of the moose, *Alces alces*. *Ecology*, **77**, 1493–500.

Sæther, B.-E., Engen, S., and Lande, R. 1996b. Density-dependence and optimal harvesting of fluctuating populations. *Oikos*, **76**, 40–6.

Sæther, B.-E., Engen, S., Islam, A., McCleery, R., and Perrins, C. 1998a. Environmental stochasticity and extinction risk in a population of a small songbird, the great tit. *American Naturalist*, **151**, 441–50.

Sæther, B.-E., Engen, S., Swenson, J. E., Bakke, Ø., and Sandegren, F. 1998b. Assessing the viability of Scandinavian brown bear, *Ursus arctos*, populations: the effects of uncertain parameter estimates. *Oikos*, **83**, 403–16.

Sæther, B.-E., Engen, S., and Lande, R. 1999. Finite metapopulation models with density-dependent migration and stochastic local dynamics. *Proceedings of the Royal Society London B*, **266**, 113–18.

Sæther, B.-E., Engen, S., Lande, R., Arcese, P., and Smith, J. N. M. 2000a. Estimating the time to extinction in an island population of song sparrows. *Proceedings of the Royal Society London B*, **267**, 621–6.

Sæther, B.-E., Tufto, J., Engen, S., Jerstad, K., Røstad, O. W., and Skåtan, J. E. 2000b. Population dynamical consequences of climate change for a small temperate songbird. *Science*, **287**, 854–6.

Sæther, B.-E., Engen, S., and Solberg, E. J. 2001. Optimal harvest of age structured populations of moose *Alces alces* in a fluctuating environment. *Wildlife Biology*, **7**, 171–9.

Sæther, B.-E., Engen, S., Lande, R., Visser, M., and Both, C. 2002a. Density dependence and stochastic variation in a newly established population of a small songbird. *Oikos*, **99**, 331–7.

Sæther, B.-E., Engen, S., and Matthysen, E. 2002b. Demographic characteristics and population dynamical patterns of solitary birds. *Science*, **295**, 2070–3.

Sanders, H. L. 1968. Marine benthic diversity: a comparative study. *American Naturalist*, **102**, 243–82.

Saurola, P. 1989. Ural Owl. In *Lifetime reproduction in birds* (ed. I. Newton), pp. 327–45. Academic Press, London.

Scheffé, H. 1959. *The analysis of variance*. Wiley, New York.

Schemske, D. W. 2002. Ecological and evolutionary perspectives on the origins of tropical diversity. In *Foundations of tropical forest biology: classic papers with commentaries* (eds R. Chazdon and T. Whitmore), pp. 163–73. University of Chicago Press, Chicago.

Schemske, D. W., Husband, B. C., Ruckelshaus, M. H., Goodwillie, C., Parker, I. M., and Bishop, J. G. 1994. Evaluating approaches to the conservation of rare and endangered plants. *Ecology*, **75**, 584–606.

Schippers, P., Verschoor, A. M., Vos, M., and Mooij, W. M. 2001. Does 'supersaturated coexistence' resolve the 'paradox of the plankton'? *Ecology Letters*, **4**, 404–7.

Schluter, D. 2000. *The ecology of adaptive radiation*. Oxford University Press, Oxford.

Schrope, M. 2001. Biologists urge US to build marine reserves. *Nature*, **409**, 971.

Schweder, T. and Holden, R. 1994. Relative abundance series for minke whales in the Barents Sea, 1952 to 1983. *Report of the International Whaling Commission*, **44**, 323–32.

Scott, J. M., Tear, T. H., and Davis, F. (eds). 1996. *Gap analysis: a landscape approach to land management issues*. American Society of Photogrametry and Remote Sensing, Bethesda, Maryland.

Seber, G. A. F. 1982. *The estimation of animal abundance and related parameters*. 2nd edn, Charles Griffin and Company, London.

Sepkoski, J. J., Jr. 1978. A kinetic model of Phanerozoic taxonomic diversity. I. Analysis of marine orders. *Paleobiology*, **4**, 223–51.

Sepkoski, J. J., Jr. 1993. Ten years in the library: new data confirm paleontological patterns. *Paleobiology*, **19**, 43–51.

Sepkoski, J. J., Jr. 1995. Patterns of Phanerozoic extinction: a perspective from global data bases. In *Global events and event stratigraphy* (ed. O. H. Walliser), pp. 35–51. Springer-Verlag, Berlin.

Shaffer, M. L. 1981. Minimum population sizes for species conservation. *BioScience*, **31**, 131–4.

Shaffer, M. L. 1983. Determining minimum viable population sizes for the grizzly bear. *International Conference on Bear Research and Management*, **5**, 133–9.

Shaffer, M. 1987. Minimum viable populations: coping with uncertainty. In *Viable populations for conservation* (ed. M. E. Soulé), pp. 11–34. Cambridge University Press, Cambridge.

Shannon, C. E. and Weaver, W. 1962. *The mathematical theory of communication*. University of Illinois Press, Urbana.

Sharpton, V. L., Dalrymple, G. B., Marin, L. E., Ryder, G., Schuraytz, B. C., and Urrutiafucugauchi, J. 1992. New links between the Chicxulub impact structure and the Cretaceous Tertiary boundary. *Nature*, **359**, 819–21.

Sharpton, V. L., Burke, K., Camargozanoguera, A., Hall, S. A., Lee, D. S., Marin, L. E., Suarezreynoso, G., Quezadamuneton, J. M., Spudis, P. D., and Urrutiafucugauchi, J. 1993. Chicxulub multiring impact basin—size and other characteristics derived from gravity analysis. *Science*, **261**, 1564–7.

Shigesada, N. and Kawasaki, K. 1997. *Biololgical invasions: theory and practice*. Oxford University Press, Oxford.

Simpson, E. H. 1949. Measurement of diversity. *Nature*, **163**, 688.

Sjögren-Gulve, P. and Ebenhard, T. (ed.). 2000. The use of population viability analyses in conservation planning. *Ecological Bulletins*, **48**, 1–203.

Skellam, J. G 1951. Random dispersal in theoretical populations. *Biometrika*, **38**, 196–218.

Smith, F. A., Betancourt, J. L., and Brown, J. H. 1995. Evolution of body-size in the woodrat over the past 25,000 years of climate change. *Science*, **270**, 2012–14.

Solberg, E. J. and Sæther, B-E. 1994. Male traits as life history variables: annual variation in body mass and antler size in moose (*Alces alces*). *Journal of Mammalogy*, **75**, 1069–79.

Solberg, E. J. Sæther, B-E., Strand, O., and Loison, A. 1999. Dynamics of a harvested moose population in a variable environment. *Journal of Animal Ecology*, **68**, 186–204.

Solow, A. R. 1993. A simple test for change in community structure. *Journal of Animal Ecology*, **62**, 191–3.

Solow, A. R. 1998. On fitting a population model in the presence of observation error. *Ecology*, **79**, 1463–6.

Soulé, M. E. 1980. Thresholds for survival: maintaining fitness and evolutionary potential. In *Conservation biology, an evolutionary-ecological perspective* (eds M. E. Soulé and B. E. Wilcox), pp. 151–70. Sinauer, Sunderland, Massachusetts.

Soulé, M. E. and Terborg, J. (ed.). 1999. *Continental conservation: scientific foundations of regional conservation networks*. Island Press, Washington, D. C.

Spellerberg, I. F. 1991. *Monitoring ecological change*. Cambridge University Press, Cambridge.

Spretke, T. 1998. Zur Prädation von Silbermöwen *Larus argentatus* bei Fluss-seeschwalben *Sterna hirundo* auf der Insel Kirr. *Vogelwelt*, **119**, 205–8.

Stanley, H. E. 1987. *Introduction to phase transitions and critical phenomena*. Oxford University Press, Oxford.

Stamps, J. A. 1991. The effect of conspecifics on habitat selection in territorial species. *Behavioral Ecology and Sociobiology*, **28**, 29–36.

Steadman, D. W. 1995. Prehistoric extinctions of Pacific Island birds: biodiversity meets zooarchaeology. *Science*, **267**, 1123–31.

Stephens, P. A., Sutherland, W. J., and Freckleton, R. P. 1999. What is the Allee effect? *Oikos*, **87**, 185–90.

Stenseth, N. C., Chan, K. S., Tong, H., Boonstra, R., Boutin, S., Krebs, C. J., Post, E., O'Donoghue, M., Yoccoz, N. G., Forchhammer, M. C., and Hurrell, J. W. 1999. Common dynamic structure of Canada lynx populations within three climatic regions. *Science*, **285**, 1071–3.

Stork, N. E. and Samways, M. J. 1995. Inventorying and monitoring. In *Global biodiversity assessment* (ed. V. H. Heywood), pp. 453–605. Cambridge University Press, Cambridge.

Strömgren, T., Lande, R., and Engen, S. 1973. Intertidal distribution of the fauna on muddy beaches in the Borgenfjord area. *Sarsia*, **53**, 49–70.

Stuart, A. and Ord, J. K. 1994. *Kendall's advanced theory of statistics*, Vol. 1, *Distribution theory*. 6th edn, Edward Arnold, London.

Stuart, A. and Ord, J. K. 1999. *Kendall's advanced theory of statistics*, Vol. 2A, *Classical Inference and the linear model*. 6th edn, Edward Arnold, London.

Stubsjøen, T., Sæther, B.-E., Solberg, E. J., Heim, M., and Rolandsen, C. M. 2000. Moose (*Alces alces*) survival in three populations in northern Norway. *Canadian Journal of Zoology*, **78**, 1822–30.

Suarez, A. V., Holway, D. A., and Case, T. J. 2001. Patterns of spread in biological invasions dominated by long-distance jump dispersal: Insights from Argentine ants. *Proceedings of the National Academy of Sciences, USA*, **98**, 1095–100.

Sugihara, G. 1980. Minimal community structure: an explanation of species abundance patterns. *American Naturalist*, **116**, 770–87.

Sutcliffe, O. L., Thomas, D. C., and Moss, D. 1996. Spatial synchrony and asynchrony in butterfly dynamics. *Journal of Animal Ecology*, **65**, 85–95.

Swenson, J. E., Wabakken, P., Sandegren, F., Bjärvall, A., Franzén, R., and Söderberg, A. 1995. The near extinction and recovery of brown bears in Scandinavia in relation to the bear management policies of Norway and Sweden. *Wildlife Biology*, **1**, 11–25.

Swenson, J. E., Sandegren, F., and Söderberg, A. 1998. Geographic expansion of an increasing bear population: evidence for presaturation dispersal. *Journal of Animal Ecology*, **67**, 819–26.

Tamarin, R. H. (ed.). 1978. *Population regulation. Benchmark Papers in Ecology*, Vol. 7. Dowden, Hutchinson & Ross, Stroudsburg, Pennsylvania.

Tanaka, T., Tokuda, K., and Kotero, S. 1970. Effects of infant loss on the interbirth interval of Japanese monkeys. *Primates*, **11**, 113–17.

Tasker, M. L., Camphuysen, M. C. J., Cooper, J., Garthe, S., Montevecci, W. A., and Blaber, S. J. M. 2000. The impacts of fishing on marine birds. *ICES Journal of Marine Science*, **57**, 531–47.

Taylor, B. L. 1995. The reliability of using population viability analysis for risk classification of species. *Conservation Biology*, **9**, 551–8.

Taylor, B. L. and Wade, P. R. 2000. Best abundance estimates and best management: why they are not the same. In *Quantitative methods for conservation biology* (eds S. Ferson and M. Burgman), pp. 96–108. Springer-Verlag, New York.

Taylor, B. L., Wade, P. R., Stehn, R. A., and Cochrane, J. F. 1996. A Bayesian approach to classification criteria for Spectacled Eider. *Ecological Applications*, **6**, 1077–89.

Tegner, M. J. and Dayton, P. K. 2000. Ecosystem effects of fishing in kelp forest communities. *ICES Journal of Marine Science*, **57**, 579–89.

Thompson, P. M. and Ollason, J. C. 2001. Lagged effects of ocean climate on flumar population dynamics. *Nature*, **413**, 417–20.

Tier, C. and Hanson, F. B. 1981. Persistence in density dependent stochastic populations. *Mathematical Biosciences*, **53**, 89–117.

Tietenberg, T. 1998. *Environmental and natural resource economics*. 4th edn, Harper Collins, New York.

Tilman, D. and Pacala, S. 1993. The maintenance of species richness in plant communities. In *Species diversity in ecological communities* (eds R. E. Ricklefs and D. Schluter), pp. 13–25. University of Chicago Press, Chicago.

Tilman, D., Reich, P. B., Knops, J., Wedin, D., Mielke, T., and Lehman, C. 2001. Diversity and productivity in a long-term grassland experiment. *Science*, **294**, 843–5.

Tufto, J., Sæther, B.-E., Engen, S., Swenson, J. E., and Sandegren, F. 1999. Harvesting strategies for conserving minimum viable populations based on World Conservation Union criteria: brown bears in Norway. *Proceedings of the Royal Society London B, Biological Sciences*, **266**, 961–7.

Tuljapurkar, S. D. 1982. Population dynamics in variable environments. II. Correlated environments, sensitivity analysis and dynamics. *Theoretical Population Biology*, **21**, 114–40.

Tuljapurkar, S. D. 1990. *Population dynamics in variable environments*. Springer-Verlag, New York.

Turchin, P. 1990. Rarity of density dependence or population regulation with lags? *Nature*, **344**, 660–3.

Turchin, P. 1995. Population regulation: old arguments and a new synthesis. In *Population dynamics* (eds N. Cappucino and P. W. Price), pp. 19–40. Academic Press, New York.

Turchin, P. 1998. *Quantitative analysis of movement*. Sinauer, Sunderland, Massachusetts.

Turchin, P. and Taylor, A. D. 1992. Complex dynamics in ecological time series. *Ecology*, **73**, 289–305.

Turelli, M. 1977. Random environments and stochastic calculus. *Theoretical Population Biology*, **12**, 140–78.

Turelli, M. 1978. A reexamination of stability in randomly varying versus deterministic environments with comments on the stochastic theory of limiting similarity. *Theoretical Population Biology*, **13**, 244–67.

Van Valen, L. 1973. A new evolutionary law. *Evolutionary Theory*, **1**, 1–30.

Van Valen, L. 1979. Taxonomic survivorship curves. *Evolutionary Theory*, **4**, 129–42.

Veit, R. R. and Lewis, M. A. 1996. Dispersal, population growth, and the Allee effect: dynamics of the house finch invasion of eastern North America. *American Naturalist*, **148**, 255–74.

Viljugein, H., Lingjærde, O. C., Stenseth, N. C., and Boyce, M. C. 2001. Spatio-temporal patterns of mink and muskrat in Canada during a quarter century. *Journal of Animal Ecology*, **70**, 671–82.

Volterra, V. 1928. Variations and fluctuations of the number of individuals in animal species living together. *Journal du Conseil International pour l'Exploration de la Mer*, **3**, 3–51. (Reprinted as Appendix in: Chapman, R. N. 1931. *Animal Ecology*. McGraw-Hill, New York.)

Volterra, V. 1931. *Leçons sur la Théorie Mathématique de la Lutte pour la Vie*. Gauthier-Villars, Paris.

Wade, P. R. 2002. Bayesian population viability analysis. In *Population viability analysis* (eds S. Beissinger and D. R. McCullough), pp. 213–38. University of Chicago Press, Chicago.

Wagner, H. H., Wildi, O., and Ewald, K. C. 2000. Additive partitioning of plant species diversity in an agricultural mosaic landscape. *Landscape Ecology*, **15**, 219–27.

Webb, C. O. and Peart, D. R. 1999. Seedling density dependence promotes coexistence of Bornean rain forest trees. *Ecology*, **80**, 2006–17.

Wei, K.-Y. and Kennett, J. P. 1983. Nonconstant extinction rates of Neogene planktonic foraminifera. *Nature*, **305**, 218–20.

Wennergren, U., Ruckelshaus, M., and Kareiva, P. 1995. The promise and limitations of spatial models in conservation biology. *Oikos*, **74**, 349–56.

White, G. C., Franklin, A. B., and Shenk, T. M. 2002. Estimating parameters of PVA models from data on marked animals. In *Population viability analysis* (eds S. Beissinger and D. R. McCullough), pp. 169–90. University of Chicago Press, Chicago.

Whittaker, R. H. 1960. Vegetation of the Siskiyou Mountains, Oregon and California. *Ecological Monographs*, **30**, 279–338.

Whittaker, R. H. 1970. *Communities and ecosystems*. Macmillan, New York.

Whittaker, R. H. 1972. Evolution and measurement of species diversity. *Taxon*, **21**, 213–51.

Whittle, P. and Horwood, J. 1995. Population extinction and optimal resource management. *Philosophical Transactions of the Royal Society B*, **350**, 179–88.

Williams, C. B. 1964. *Patterns in the balance of nature and related problems in quantitative ecology*. Academic Press, London.

Williams, D. W. and Liebhold, A. M. 1995. Detection of delayed density dependence: effects of autocorrelation in an exogenous factor. *Ecology*, **76**, 1005–8.

Winkel, W. 1996. Der Braunschweiger Höhlenbrüterprogramm des Instituts für Vogelforschung Vogelwarte Helgoland. *Vogelwelt*, **117**, 269–75.

With, K. and King, A. W. 1999. Extinction thresholds for species in fractal landscapes. *Conservation Biology*, **13**, 314–26.

Wolff, J. O. 1997. Population regulation in mammals: an evolutionary perspective. *Journal of Animal Ecology*, **66**, 1–13.

Wright, S. 1931. Evolution in Mendelian populations. *Genetics*, **16**, 97–159.

Wright, S. 1940. Breeding structure of populations in relation to speciation. *American Naturalist*, **74**, 232–48.

Wright, S. J. 2002. Plant diversity in tropical forests: a review of mechanisms of species coexistence. *Oecologia*, **130**, 1–14.

Young, T. P. 1994. Natural die-offs of large mammals: implications for conservation. *Conservation Biology*, **8**, 410–18.

Zachos, J., Pagani, M., Sloan, L, Thomas, E., and Billups, K. 2001. Trends, rhythms, and aberrations in global climate 65 Ma to present. *Science*, **292**, 686–93.

Zeng, Z., Nowierski, R. M., Taper, M. L., Dennis, B., and Kemp, W. P. 1998. Complex population dynamics in the real world: modeling the influence of time-varying parameters and time lags. *Ecology*, **79**, 2193–209.

Index